Mending **Manufacturing**

How America Can Manufacture Its Survival

PEGGY SMEDLEY

Specialty Publishing Company

Cataloging-in-Publication Data is on file with the Library of Congress.

Printed in the United States of America
First Edition

Smedley, Peggy 2004
Mending Manufacturing
How America Can Manufacture Its Survival
Peggy Smedley
216 p.: ill
Bibliographical references and index

ISBN 0-9755199-0-5

1. Manufacturing 2. Manufacturing Industries United States
3. Economic Policy United States
I. Title.
10 9 8 7 6 5 4 3 2 1

Specialty Publishing Company, Inc.
135 E. St. Charles Rd., Carol Stream, Il 60188
www.specialtypub.com

Table of Contents

List of Figures **iii**
Guide to Abbreviations **v**
Acknowledgments **vii**
Foreword **1**

Chapter 1
Facing the Crisis **13**
Chapter 2
Economic Recovery **37**
Chapter 3
Reversing the Trend **47**
Chapter 4
Why Is Manufacturing Important? **55**
Chapter 5
Globalization and Innovation **55**
Chapter 6
Action Plan **71**
Chapter 7
Fighting for Change **83**
Chapter 8
Manufacturers Speak Out
DWIGHT CAREY
American Productivity Design
and Equipment Inc. **93**
JACK BOLICK
Honeywell Process Solutions **105**

GUS WHALEN

Warren Featherbone Co. 115

BRETT KINGSTONE

Super Vision International 122

JOSEPH RAGOSTA

Oseco, Inc. 128

HOWARD SANDERS

EMF Corporation 134

Chapter 9

Politicians Speak

RON PAUL

Rep. (R-Texas) 139

GRANT ALDONAS

U.S. Department of Commerce 145

MURRAY SABRIN

Professor and former Political Candidate 162

DONALD MANZULLO

Rep. (R-Ill.) 171

MARCY KAPTUR

Rep. (D-Ohio.) 177

Chapter 10

The Mending Has Begun 183

References 199

Index 203

List of Figures

Figure 1.1	Imports (Goods)	18
Figure 1.2	Employment	21
Figure 1.3	Manufacturing Job Decline Breakdown	22
Figure 1.4	U.S. Productivity and Cost Index	24
Figure 1.5	Clinton Administration	25
Figure 1.6	Bush Administration	25
Figure 1.7	Manufacturing Job Decline	27
Figure 2.1	PMI Index	38
Figure 2.2	PMI Index 2004	40
Figure 2.3	Manufacturing Output	41
Figure 2.4	Global Manufacturing PMI	44
Figure 5.1	Federal Research and Development Spending	65
Figure 6.1	U.S. Unemployment	76
Figure 7.1	GDP by Industry	86
Figure 10.1	U.S. Economic Growth	185
Figure 10.2	Exports 2002	193

Guide to Abbreviations

AAMA	American Apparel Manufacturers Association
AMT	Alternative Minimum Tax
CEO	Chief Executive Officer
CHIP	Computer Hacking and Intellectual Property
CRM	Customer-Relationship Management
DOJ	Department of Justice
ERP	Enterprise Resource Planning
EU	European Union
FCS	Foreign and Commercial Service
FRBSF	The Federal Reserve Bank of San Francisco
FTAA	Free Trade Area of the Americas
GATT	General Agreement on Tariffs and Trade
GDP	Gross Domestic Product
GOP	Grand Old Party (U.S. Republican Party)
HEM	Honeywell Electronic Materials
IC	Integrated Circuits
ICT	Information and Communication Technology
IFPMM	International Federation of Purchasing and Materials Management
IMF	International Monetary Fund
IPR	Intellectual Property Rights
IRS	Internal Revenue Service
ISM	Institute for Supply Management
ISO	International Organization for Standardization
ITC	International Trade Commission
LED	Light Emitting Diode
LLC	Limited Liability Company
MEP	Manufacturing Extension Partnership

MERC	Manufacturing and Economic Recovery Conference
MES	Manufacturing Execution System
MRP	Materials Requirements Planning
NAFTA	North American Free Trade Agreement
NIST	National Institute of Standards and Technology
NTB	Non-Tariff Barriers
OEM	Original-Equipment Manufacturer
OPEC	Organization of the Petroleum Exporting Countries
PLM	Product-Lifecycle Management
PMI	Purchasing Manager's Index
QR2	Quick Response Applied Across Companies
R&D	Research and Development
ROI	Return on Investment
SBA	Small Business Administration
SBIC	Small Business investment Companies
SCM	Supply-Chain Management
SEC	Securities and Exchange Commission
SIA	Semiconductor Industry Association
SMEs	Small and Medium Enterprises
SMM	Small and Medium Manufacturers
SRM	Supplier-Relationship Management
TC2	Textile/Clothing Technology Corporation
TQM	Total Quality Management
UAW	United Autoworkers (Union)
USBIC	United States Business and Industry Council
USTR	United States Trade Representative
VAT	Value-Added Tax
WTO	World Trade Organization

Acknowledgments

Typically, the final words that are put on the pages of a book are the acknowledgments. Although they are the last words to be put in a book, they are the ones that really should go first. With that said, I would like to express my sincere thanks to all the people who have inspired me to write this book and to those who took the time to keep me on the straight and narrow when completing this lofty endeavor.

It is very hard for me to put into words the appreciation and gratitude I have for all the people who contributed long hours to make this book a reality. No book is ever a solo performance, and this book is by no means an exception to that rule. Many people have spent countless hours shaping its content and direction.

I'm especially grateful to Dwight Carey, Jack Bolick, Gus Whalen, Brett Kingstone, Howard Sanders, Joseph Ragosta, Congresswoman Marcy Kaptur, Congressman

Ron Paul, Congressman Donald Manzullo, Grant Aldonas, and Murray Sabrin for graciously sharing and conveying their personal and professional thoughts about the challenges facing the manufacturing industry within the United States.

Many thanks to Alan Tonelson for taking the time to write the foreword to the very complicated and controversial subject this book tackles.

I also want to give special mention to Mike Collins for holding my feet to the fire and making me think about every word that was put on each and every page. His cooperation and critical eye made it possible for me to address a very difficult topic.

I'm very proud and honored to have worked with all these people.

I would be remiss if I did not single out a couple of people who I'm very proud to work with on a daily basis.

Working side by side for many years, John Buell continues to be essential in shaping the vision and direction of our efforts to cover the manufacturing sector. No other person has been in the thick of the manufacturing crisis with me more than John. John continues to be a much-needed confidant. He constantly reminded me of what needed to be identified to communicate a strong and cohesive message. He was an advocate for *Mending Manufacturing* from the very beginning, recognizing its importance to the marketplace.

Mark Emond and I go a long way back and I don't believe there is another writer on this plan that I respect more. Mark has been enthusiastic and supportive offering timely words of encouragement when I needed it the most. I also want to recognize Jean Emond. Her exceptional copyediting proved pivotal in completing this Herculean task. No one could ever ask for a better supporting cast.

Thanks also to Michael Jarosik and Marco Carating for their hard work and assistance on this project.

Last, but certainly not least, I would like to give a special thank you to my family. I want to thank my father, Bill Spathies, my sister BJ, and my brothers for their ongoing support. Most of all I especially want to thank my husband, Dave, and my children—Christina, David, and Aaron—who have been such a source of support and encouragement. Writing this book was not an easy task, but I could not have completed it without the strength of my husband whose constant support proved to be my inspiration.

And finally, I would like to thank the myriad of *Start* magazine readers who continue to share their experiences and wisdom. They have all played a pivotal role in making this all possible.

While this book only touches the surface of the many issues impacting manufacturing, it goes a long way in demonstrating what we all have accomplished and what we can accomplish if we continue to work together.

In loving memory of my mother Josephine who instilled in me nothing happens without a dream.

Foreword

Intellectual Ruts

To steal a line from the 1970s film, "Broadcast News," American domestic manufacturers are mad as hell, and they're not going to take it anymore.

Faced with what looks like the worst crisis in their collective history, a large and steadily growing percentage of companies that develop and make the lion's share of their products in the United States have started to mobilize politically. Their message to political leaders: Help us fight back.

Mending Manufacturing is a leading example of the new activism in domestic manufacturing. This book represents a key step in an ongoing campaign to identify the main roots of the manufacturing crisis, spread the word among the nation's industrial community, and develop not only new ideas for reviving this critical industrial sector, but the political strategies to make such measures reality.

Judging from the book, from the conferences that *Start* magazine has sponsored, and from other recent outbursts of grassroots activity by domestic manufacturers, America's industrialists are forming an even clearer picture of the challenges confronting them. They understand that today's manufacturing crisis has much deeper and more fundamental causes than the latest economic downturn, and that the recent uptick in economic and manufacturing activity can no longer obscure the sector's grave structural problems. They recognize that, for all the valid complaints about high business costs, domestic manufacturers' main problems cannot be solved without wholly new international trade and other globalization policies. Finally, they increasingly acknowledge that the interests of multinational manufacturers on the one hand, and the smaller domestic manufacturers still prominent in their production chains on the other, have greatly diverged in the last two decades.

Nonetheless, domestic manufacturers still have some ways to go in fully comprehending the crisis threatening to engulf them, and thus in devising the most effective ideas for overcoming it. As a result, the educational process launched by Peggy Smedley in conjunction with *Start* magazine and so many other organizations must be continued and intensified.

THE NATURE OF THE CRISIS

As *Mending Manufacturing* astutely observes, although the conventional economic figures insist that a strong manufacturing comeback has begun, contrary anecdotal evidence is too widespread to be ignored. Particularly striking is the number of domestic manufacturers who assert that their current predicament is the most serious in the history of their companies. These claims deserve special attention for two main reasons.

First, many domestic manufacturing companies are family-owned businesses whose ownership bloodlines have remained unchanged for decades and even longer. The executive ranks of these companies represent a uniquely credible stock of historical memory. The contemporary CEOs who warn of unprecedented threats to their companies and industries are using frames of reference created by personal experience or first-hand accounts. Their sources are literally their own fathers and grandfathers.

Second, many of today's domestic manufacturers are companies with long records of withstanding the hardest economic times and the strongest competitors—including the Great Depression of the 1930s and the Japanese export surges of the 1970s and 1980s. To survive, they truly needed to be the fittest, constantly improvising, adapting, and innovating. So when their leaders warn that today's challenges may be fatal without major policy course corrections, the nation should pay attention. These aren't statements coming from chronic whiners.

Nonetheless, statistical bases for their claims do exist—if one is curious enough to look for them. For example, although Washington does not keep these figures as such, raw output and trade data make possible the calculation of import penetration rates for hundreds of key industries. These trends in marketshare are in many ways a more revealing indicator of competitiveness than the trade deficit numbers. And they show that from 1992 to 2001 (the last year for which comprehensive data are available), about 90% of America's most important manufacturing industries lost significant marketshare to goods produced overseas.

In addition, as widely noted, the current economic recovery is probably built on sand. Interest rates remain at multidecade lows, tax rates have fallen steeply, federal deficits are at record absolute levels, and the dollar keeps drifting down against most major currencies. Yet despite this unprecedented multidimensional stimulus program, growth rates recently have been quite ordinary by historical standards. And these particular engines of growth are anything but long-lasting. When the fuel runs out, the long-standing ills of the domestic manufacturing sector—and of the larger economy—will be exposed more starkly than ever.

THE CRITICAL TRADE ANGLE

The current administration and the nation's major business lobbies recognize that something is seriously wrong with U.S.-based manufacturing, but their explanation should satisfy no one. It's true that the costs of doing business in the United States are higher than the costs of doing business in much of the rest of the world, and that many reductions are possible and desirable—especially in the healthcare field. But it is deeply misleading to portray the tax and regulatory burdens borne by domestic manufacturers as the main source of their troubles.

In the first place, these costs collectively represent a fundamental decision made by the American political system decades ago and strongly upheld ever since to develop a mixed system of capitalism with substantial social and economic protections. In other words, starting in the 1930s (and before, in some instances), the United States decided to become a First World country.

And although the exact scale and makeup of the resulting welfare state can and should be continually debated, its basic dimensions are here to stay, and should in fact be a source of great national pride. Indeed, the implicit assumption that global competitiveness requires America to revert to the social Darwinist doctrines of the 19th century, and to surrender the gains made since the New Deal era, is not only politically impractical, it is morally unacceptable as well.

Moreover, the focus on domestic costs overlooks the reality that when it comes to the multinational companies that have sent so many jobs and so much production overseas, the oft-discussed corporate tax burden is largely a chimera.

In the best of all possible worlds, the United States would shift to some version of a value-added tax (VAT), which with the proper carve-outs for low-income Americans, would restore a sustainable balance in the U.S. economy between consumption and production. But politically, a VAT-like scheme has long been a nonstarter.

Nonetheless, for all the complaints about the corporate income tax and the allegedly outsourcing-friendly provisions of the U.S. tax code, the inarguable reality is that few multinational companies pay significant corporate taxes. As recent reports by the General Accounting Office and other reputable organizations demonstrate, the armies of expensive paid accountants and tax lawyers employed by these firms to reduce and even eliminate tax liabilities have more than earned their pay.

In addition, even nominal tax rates are hardly the only or even the most significant cost differential between doing business in the United Sates and doing business in a country like China, destination of so much U.S. outward-bound direct investment. The latter, after all, provides not only major tax breaks for such investment, it also showers companies with subsidies for land, water, raw materials, and other key industrial inputs, and of course manipulates its exchange rate, rarely enforces environmental laws, and drives down rockbottom labor costs yet further through universal and often violent repression of worker rights.

Trying to match Chinese or Mexican or Indian business-cost levels would plunge the U.S. economy into a real race to the bottom in living standards that would impoverish the American people and do virtually nothing to enrich Third Worlders.

The key to maintaining U.S. industrial competitiveness as well as First World living standards is to recognize how current U.S. trade and globalization policies are undermining the domestic manufacturing base and its ability to offer good wages and benefits to its workers.

Trade policy critics have pinpointed some elements of the problem; e.g., the decades-long failure by Washington to open foreign markets adequately and to enforce trade agreements already on the books. Yet these failures are only symptoms of the real problem with U.S. trade policy: Since the early 1990s, its basic purpose has been to encourage multinational companies to send production facilities from the United States to low-income countries like China, and to use these countries as new bases for supplying the U.S. market.

The best evidence for this claim is the overwhelming Third World focus characterizing U.S. trade policy since the late 1980s. Starting with the first proposals to bring Mexico into the U.S.-Canada Free Trade Agreement, there can be no doubt that American trade policymakers have worked much harder to sign trade deals with Third World countries than with Japan and its fully industrialized counterparts in Western Europe.

During the early Clinton years, this focus was explained by describing Third World countries as emerging markets that were rapidly turning capitalistic and that would surely register growth rates and create new export opportunities much faster than would the world's mature developed economies. Yet even before a series of financial crises in the mid and late 1990s turned many of these emerging markets into vanishing markets, the numbers never supported this view.

At the time of NAFTA's signing, for example, the Mexican economy was about one-thirtieth the size of the U.S. economy. How could it possibly consume enough U.S.-made products to be a significant driver of U.S. growth? Throughout the 1990s, wages in China were falling behind the rate of inflation. How could even a billion people become major purchasers of U.S. products when the vast majority were desperately poor and becoming poorer? More recently, Morocco and six Central American countries have dominated the U.S. trade policy agenda. Yet as of 2002, the former represented a market for American goods only slightly larger than that of New Haven, Conn. The latter was smaller economically than greater West Palm Beach, Fla.

U.S. multinational companies plainly have been aware of all these realities. Rather than valuing Third World trade partners as consumer markets, they valued them as sources of low-wage workers and as easy regulatory environments that they could turn into export platforms. They recognized that, once the glitches of doing business in the Third World were overcome, expanded trade relations with these countries could give them a world in which they could pay their workers Chinese or Indian wages, and yet charge their customers (who were mainly in the United States) American prices.

Thus, the challenges presented by China and the rest of the world (and particularly the Third World) to U.S. domestic manufacturing are not mainly the outgrowth of autonomous progress in these countries or of some allegedly natural tendency of technology to diffuse from its source. Rather, these challenges result from U.S. (and, increasingly, other countries' multinationals) transferring technology and other forms of manufacturing know-how to low-income countries right after they develop it in the home country.

As a result, those who warn domestic manufacturers against bashing foreigners are to a great extent correct. Although foreign mercantilist practices ranging from illegal subsidies to intellectual property theft do plague U.S. domestic manufacturing, the main culprits are U.S. companies themselves and the outsourcing-centered trade strategies they have spearheaded. But the fact that so much of "the enemy...is us" doesn't mitigate the damage these globalization policies are doing to the domestic manufacturing base and by extension to America's national security and future economic prospects.

By imposing on the nation trade policies that guarantee low-cost foreign production sites virtually unfettered access to the high-cost U.S. market, multinational companies have forced their domestic supply base into a competition that a great majority of these companies can't win—at least without driving down U.S. wages and benefits still further. This competition, moreover, will not only continue as long as the access to the U.S. market exists, it will metastasize throughout the supply chains of a rapidly growing number of industries.

The consequences of this expanding competition could not be clearer. In the short term, American multinational companies will enjoy wider profits and American consumers will enjoy lower-priced goods and a greater variety of goods. Over the longer term, however, the multinationals will increasingly find themselves short of customers able to pay for their purchases responsibly. The Americans they fire will need to sink deeper and deeper into debt to continue their present rate of consumption, and the low wages they pay abroad will prevent their foreign workforces from filling the gap.

On an economywide level, these trends will produce an America ever more reliant on foreign borrowing to maintain anything close to current living standards, but a world ever more reliant on this debt-strapped America for its growth. Good luck writing a happy ending to this story.

OBSTACLES TO SUCCESS

Yet however clear the danger, even American domestic manufacturers who do understand them will face major obstacles in their struggle for trade policy reform. The first and biggest obstacle is the multinational manufacturing community whose short-term interests are so powerfully served by today's outsourcing-centered trade strategy. These companies not only dominate political Washington with their lavish campaign contributions to both parties but they dominate intellectual Washington as well with their comparably important support for the so-called independent think tanks.

Two other obstacles are much less obvious, but nearly as important. Mainstream liberals are finally expressing fairly uniform opposition to the basic trade agreements recipe of the 1990s. Partly because the Democrats lost the White House in 2000 and partly because of the current manufacturing crisis and the weaker economy, even many liberals who supported the NAFTA-like trade deals of the Clinton era are now calling for new approaches.

Unfortunately, their recommendations will have only symbolic effect at best. Inserting into new trade agreements strong, allegedly enforceable provisions safeguarding worker rights and environmental protection in trade-partner countries can't possibly reduce the flight from the United States of manufacturing capacity and jobs to any significant degree. As documented in my recent book, *The Race to the Bottom* and elsewhere, even if Third World countries created the most union-friendly labor-rights regimes imaginable, the enormous oversupply of labor in these countries will keep wages abysmally low for the indefinite future. Moreover, given how difficult it is to enforce labor and environmental laws in the United States, it strains credulity to assert that meaningful enforcement is possible abroad.

Until mainstream liberals recognize that much more dramatic changes in U.S. trade policy are needed, their focus on labor and environmental issues will continue to keep the national trade policy debate focused on sideshows, and valuable time and resources needed to revive manufacturing will be wasted.

Domestic manufacturers also need to deal with the related problems posed by the Third World advocacy groups that have become so prominent in the politics of U.S. and world trade. Organizations like Oxfam International and various progressive faith-based charities have played a highly influential role in world trade affairs since the successful protests they helped stage at the 1999 World Trade Organization meeting in Seattle.

Although they have worked closely with the American

labor movement, the main effect of their activities has been not to highlight the damage being done to American workers by current globalization policies, but the damage being done to Third World countries by these policies. And that's exactly what these groups want. As a result, much of the globalization debate even inside the United States has changed from arguments over whether globalization has been helped or harmed Americans to whether it has helped or harmed the developing countries.

Consequently, it should not be surprising that the current round of WTO world-trade talks is explicitly aiming to secure more benefits for Third World countries than for the rest of the world, including the United States. The interests of American workers and American domestic companies alike have been completely shunted aside.

CONCLUSION

America's domestic manufacturers are off to a strong start in awakening the nation to the gravity of the manufacturing crisis. But now they need to take some daunting next steps. Of course, they need to devote significant resources to the pro-manufacturing campaign, and to ensure that these funds are wisely spent. Domestic manufacturers are unlikely to be able to match their multinational adversaries dollar-for-dollar or anywhere close. But they won't have to if they avoid duplication and other forms of waste.

They'll stretch their dollars much further if they stop subsidizing their opposition by continuing to pay dues to the main multinational-headed business lobbies. Severing longstanding connections with these organizations and the powerful companies that control them will be difficult decisions, especially since the multinationals and the domestic companies agree on most purely domestic issues. But companies that have been driven out of existence by ill-conceived trade policies won't have taxes or regulations or labor unions to worry about anyway, and that's what the multinationals' trade policy positions will inevitably produce.

Domestic manufacturers also need to start differentiating their message from that of the Third World advocacy groups and even the mainstream liberal groups. Unless they spread awareness of a more nationally focused basis for opposing today's globalization policies (not to mention an infinitely more pragmatic one), American leaders will continue wasting time debating trivialities while the domestic manufacturing base moves closer to irretrievable decline.

The only alternative is for domestic manufacturers to stay in the same organizational and intellectual ruts, in effect waiting for a nationwide economic crisis to spark public support for needed change. Tragically, of course, by then it may be too late to turn the tide. For the nation's sake and their own, America's domestic manufacturers must do better.

Alan Tonelson
Research Fellow
United States Business and Industry Council
and Author of *Race to the Bottom*

1

Facing the Crisis

The United States is facing a manufacturing emergency. Absent strong federal leadership, the hemorrhaging of factory jobs will continue to greatly hinder the economic health and prosperity of the nation. Unless Washington acts swiftly and decisively, the U.S. could plummet into a second-class industrial power. Thus, domestic producers and consumers alike need to recognize the problems hampering the United States and help to correct them before it affects the well-being of the next generation of Americans.

While these aforementioned statements might appear to be political rhetoric to most Americans, the manufacturing crisis is undermining the livelihoods of American working families and has serious consequences for the nation's economy.

At first blush, most officials in Washington, consumers, and even some manufacturers themselves do not fully grasp nor understand the economic consequences that will alter our future economic stability.

The economic outlook for the manufacturing industry hasn't been the brightest during the first few years of the 21st century. Pressures have been mounting about the migration of jobs overseas and the role China and other countries are playing in the decline of American manufacturing. Debate has been intense over the causes and solutions among manufacturers, politicians, and industry observers. Frustrated by lack of consensus and energy focused on the needs of the manufacturing industry, some U.S. producers have become even more vocal in recent years, criticizing not only the current administration, but just about every politician—Republican or Democrat—who is not working toward saving manufacturing jobs.

Republican + Democrat =
SAVING MANUFACTURING JOBS

The focus of most Americans has been on the Iraq war, when it should be on rebuilding and even reviving our economic well-being, first with an eye on manufacturing. No presidential candidate seems to understand how much manufacturing will impact national security and that the future economic health of the country is at risk. Only recently has Washington admitted there is a problem.

The president needs to be more determined to do more to help small and medium-size manufacturers find new ways to compete, although some of Bush's tax exemptions have helped. On the other hand, this is not to say that Democrat John Kerry has a better understanding of the issues plaguing manufacturing. Kerry has communicated nothing worth noting or even serious when it comes to trade, jobs, and even wages. In fact, neither the Republicans nor the

Democrats have truly grasped the enormity of a problem that began more than a century ago. While politicians have tried to voice their concerns recently, only a handful have tried to make a real difference.

Washington continues to significantly underestimate the major economic problems that need to be solved for the long-term material health of the nation. Frankly, unless Congress acts accordingly, we could be watching a powerful nation be brought to its knees. Alongside of the government, each and every manufacturer needs to step up and work in tandem to begin mending manufacturing within the United States.

What's more, American consumers, much like corporate America, needs to recognize their culpability in creating the crisis in manufacturing. Warnings about the crisis engulfing American manufacturing have worsened so much that even the media and Washington are beginning to take notice. Even though the news has been intensifying in recent months, Congress and the administration continue to come up short and have not created a plan that can be acted upon today to resuscitate manufacturing.

If the administration continues to downplay the industrial sector's woes, the health of manufacturing could be nearing the point of irreversible danger with catastrophic consequences. Sadly, as this crisis continues to gain almost unstoppable momentum, it could dramatically hinder the ability of the United States to continue to be a world leader.

As noted earlier, it's not surprising then that as the 2004 election season kicked into gear the first order of business for both the Democrats and the Republicans has been to solve the manufacturing crisis. Politicians, economists, and industry pundits continue to wrangle over how to handle the rising tide of concern and they publicly agreed to do whatever is necessary to put manufacturing back on the U.S. priority list. Honorable as their intentions might seem, politicians need to do more than pay lip service to the manufacturing crisis if they really want to solve the problems facing manufacturers.

When the Republican party gained control of the House, Senate, and the White House for the first time in 50 years, conventional wisdom was that the business-friendly Republicans would give this struggling economy a much-needed shot in the arm across multiple industries, with manufacturing looking to be one of the first beneficiaries.

With victories in the November 2002 elections, the GOP is holding a seat advantage of 51-48 in the Senate and 228-210 in the House, each holding an independent as well. Add that to the half a percentage point cut by Federal Reserve Board Chairman Alan Greenspan a day after the elections and all signs seemed to point toward the road to recovery for business conditions.

Concerns over the war with Iraq and terrorist threats of even more attacks have only increased economic pressures as the presidential election in November nears. Couple all of these issues with the rising costs of oil and gasoline prices and you have all the ingredients for yet another economic disaster.

As uncertainty intensified in recent months, the Commerce Department issued a report called, "Manufacturing in America: A comprehensive Strategy to Address the Challenges to U.S. Manufacturing." The report's goal was to find new ways to resolve the problems plaguing U.S. manufacturers, "and to develop a strategy designed to ensure that the government is doing all it can to create the conditions necessary to foster U.S. competitiveness in manufacturing and stronger economic growth at home and abroad."[1]

Secretary of Commerce Donald Evans admits that the burdens placed on American firms by regulations, litigation, taxes, healthcare, and expensive energy have all increased the challenges for manufacturers. Statements such as modernize the U.S. legal system and lower healthcare costs, while worthy on the surface, will not be implemented within the necessary timeframe to have an impact on domestic American manufacturers.

After months of work and much industry review, it's apparent the Bush administration still lacks a clear understanding of what is ailing American manufacturing. Even the United States Business and Industry Council (USBIC) has been harsh in its criticism of the administration, stressing that when it came to presenting remedies, the report set sweeping and unachievable goals. The USBIC insists the White House continues to come up short and believes that much stronger measures than those proposed in the report are needed to keep American industry strong in the face of surging foreign imports.[2]

Many manufacturers insist that to maintain American prosperity, security, and global leadership, much needs to be done to effectively address the challenge from foreign competitors—whose home markets are protected and even subsidized.

On the surface it appears as if the administration and its officials are following the party line with their speeches, which is that the economy is getting better and jobs will be created, unemployment is going down, and inflation is under control. Only in recent months have we been witnessing an uptick in manufacturing employment. Still, many argue that there are facts contrary to all of these comments and with all of the political promising and the dollar issue, and not facing up to the terrible cost of regulation, many manufacturers are not optimistic that Washington will resolve the crisis any time soon.

What's more, job creation in the United States is a heavily partisan issue fraught with potentially serious repercussions if it turns out that strong recovery isn't in the cards. Since the largest manufacturers are outsourcing and shifting much of their production overseas, will small businesses hire more and help fill the shortfall?

Figure 1.1 **IMPORTS (GOODS)**

Data are goods only, on a Census Basis, in billions of dollars

	Imports (Year-to-Date)	Rank	Percent of Total Trade
Total, All Countries	568.1	–	100%
Total, Top 15 Countries	436.7	–	76.9
Canada	103.0	1	18.1
China	68.9	2	12.1
Mexico	61.9	3	10.9
Japan	52.3	4	9.2
Germany	31.0	5	5.5
United Kingdom	19.1	6	3.4
South Korea	17.9	7	3.1
Taiwan	13.4	8	2.4
France	12.2	9	2.1
Ireland	11.7	10	2.1
Italy	11.0	11	1.9
Malaysia	10.6	12	1.9
Venezuela	9.6	13	1.7
Brazil	7.2	14	1.3
Saudi Arabia	6.8	15	1.2

Source: FTD WebMaster, Foreign Trade Division, U.S. Census Bureau

It cannot be overstated that the industrial sector is suffering. And as much as most Americans cannot even imagine that domestic manufacturers are suffering, these facts speak for themselves. Imports have taken $1.2 trillion worth of domestic sales, and the trade deficit is expected to soar to a whopping $500 billion.[3]

Clearly, the big economic issue for this year's presidential election will continue to be job growth. Many of the nation's largest corporations are sending jobs abroad as a business strategy meant to reduce labor costs and improve profits.

It appears so far that most of the job growth has been dependent on small business. The question that arises if the economy does not make a full recovery is whether small businesses can continue to carry the burden and spark the economy back to health. Also, can domestic small and medium-size manufacturers produce enough to offset the loss in jobs to overseas operations that America's largest corporations are making?

A key finding of the latest Wells Fargo/Gallup Small Business Index says the reason job creation in the United States has been so anemic might be that some individual small businesses are laying off at about the same rate as they are hiring, and also, of course, the fact that large corporations are reducing their numbers of employees in the United States.

Normally, small businesses account for more than half of the country's private-sector jobs, and many more small-business owners plan to increase, rather than decrease, their hiring during the next 12 months—a positive sign for job creation.

On the other hand, the strong economic expansion during the past three quarters should have meant job growth by small businesses by the end of the year. The question that remains is will small business, including manufacturers, continue to expand or will small business owners be discouraged from hiring new employees? If so, then the new-job-creation problem will continue to be much more serious than previously anticipated.

A bigger problem is arising because small and midsize manufacturers are finding it harder and harder to compete. To date the large manufacturers have been beating their smaller suppliers to submission, asking them to compete with the low-cost labor being created in China.

Small and midsize manufacturers are suffering at the hands of bullies. While most Americans are content with

purchasing the cheapest products, what they do not realize is this comes at the expense of American manufacturers and American workers. This type of treatment is no different than the schoolyard bully we all tried to avoid in grammar school. While this intense pressure is enabling goods to be produced at lower costs, at the same time it is squeezing the smaller firms out of business. Ultimately, this approach will squeeze every dollar out of their suppliers and they will certainly be the ones who will suffer greater in the end. We are beginning to see this happening in the steel market with the demands being created by China for more steel. China's rapid industrialization has significantly increased its use of steel, and as this use grows, the demand for steel grows worldwide, causing price increases and steel shortages.

While manufacturers would like to walk away from the lessons we learned as children, unfortunately this does not hold true today. Today, larger manufacturers are ending relationships with partners because all too often someone else is offering a cheaper price for the same widget. In this case the cheaper labor is coming from China. Unfortunately, cost-cutting has become a way of life for many companies, forcing even more manufacturers to reduce their workforces, ultimately putting more and more of our neighbors out of work.

Figure 1.2 **EMPLOYMENT**

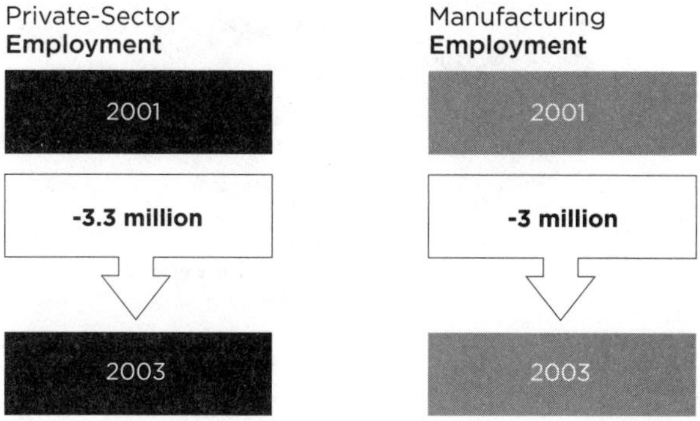

Private-Sector **Employment**	Manufacturing **Employment**
2001	2001
-3.3 million	**-3 million**
2003	2003

President Bush has repeatedly expressed concern for the unemployed—since January 2001, the United States has lost 3.3 million private-sector jobs, including nearly 3 million of them in manufacturing alone by the end of 2003.

Many conservatives insist the precipitous decline in manufacturing jobs is simply the natural result of market forces and technological change. On the other hand, many liberals blame trade and unfair foreign competition. However, none of these explanations paints a complete picture for the stagnant economy and overall loss of jobs. Failed U.S. trade policies and unabated corporate shifts of U.S. jobs to low-wage countries have exacted a very heavy toll on the manufacturing sector.

So far during the Bush administration, higher factory productivity accounts for 40% of the decline in manufacturing jobs, while slower growth in the demand for goods (as

Figure 1.3 **MANUFACTURING JOB DECLINE BREAKDOWN**

A breakdown of the causes in job decline during the Bush administration.

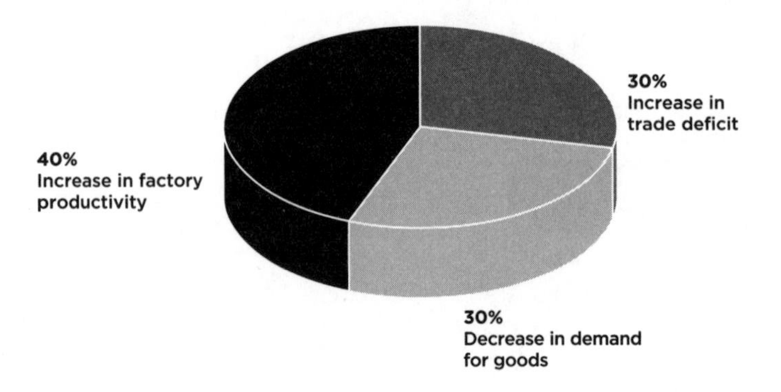

compared to services) caused almost a one-third dip. The remaining 30% of the decline in jobs was created by an increase in the trade deficit. While the job losses are significant it should be stated that the government should not intervene to prevent job losses from higher productivity. However, since a good majority of the job losses under the current administration are unrelated to productivity, it is reasonable to focus on the kinds of action steps the government needs to make to help manufacturing. It should also be noted that according to government reports, about one quarter of the decline in manufacturing jobs during the previous administration came from trade, and much of this was in lower-skilled consumer-goods industries, such as textiles and apparel, that compete largely on cost.

As U.S. firms continue to export even more jobs in manufacturing, there is even greater danger that this industry could be collapsing. Many companies are seeking cheaper ways to produce and make goods. Thus, companies are looking to China to produce less expensive parts. As a result more and more smaller manufacturers are finding it harder to compete. As a result, manufacturers are quickly becoming the victims of a national crisis that is dramatically altering the economic well-being of the American economy.

The United States added nearly half a million manufacturing jobs between 1993 and 2001, while in contrast nearly 3 million jobs have vanished since then. However, some 270,000 manufacturing jobs have been gained since the beginning of 2004.[4]

The government needs to do more to educate the American public about the facts that have contributed to the fallout that will continue if manufacturers do not begin to get their plants humming again.

Despite the rise and fall of manufacturing jobs, the current administration is not the principal culprit for the decline in American manufacturing. The problems that have occurred in manufacturing began long before the Bush administration gained control of the White House. It's not surprising, then, that the lack of government policies to address the crisis in manufacturing has confronted American workers for more than 20 years.

The biggest misconception that exists today is that the success of manufacturing is contingent upon the number of factory jobs. Fact: Advances in technology have increased productivity, leading to fewer jobs as companies produce more with fewer workers.

Another misconception is that increased productivity is really the cause for the ongoing decline in manufacturing employment. Fact: While increased productivity has reduced factory jobs, it is only partially responsible for the decline in manufacturing jobs that now exists today.

Many in Washington argue the point that the decline in manufacturing employment is similar to what occurred in agriculture during the previous century. As productivity continued to grow, workers sought out other higher paying jobs in many other segments of the economy. This is also open to some interpretation.

Figure 1.4 **U.S. PRODUCTIVITY AND COST INDEX**

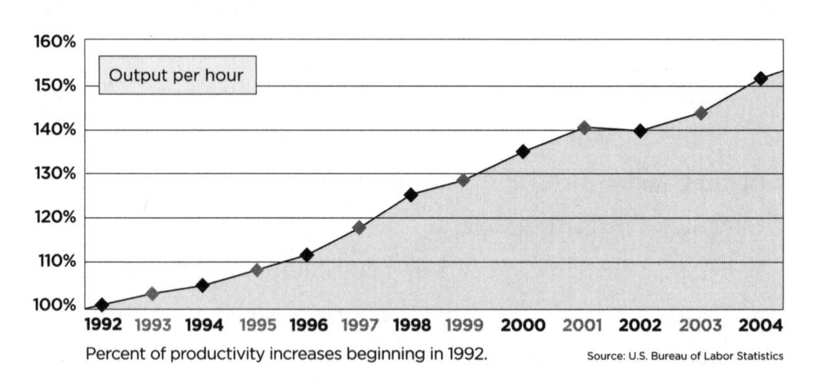

Percent of productivity increases beginning in 1992.

So let's look at the numbers. From 1992 to the end of 2003, manufacturing productivity increased 54.6%, compared with around 29% in the nonmanufacturing business sector. As a result, manufacturers were able to produce more with fewer workers relative to the rest of the economy.[5]

During his eight-year term in office, President Bill Clinton was able to see overall jobs grow by 8.6% to 138 million and manufacturing employment increase by about 332,000 jobs but only at a 1% pace. During the same period, manufacturing jobs rose to a peak of 17.7 million in June 1998, but slowly saw a decline of nearly 1 million manufacturing jobs, dropping to 16.9 million when he left office in January 2001.[6]

Interestingly, during the Bush administration, the loss of 1.4 million jobs is said to have nothing to do with overall productivity, and more to do with what Americans might be consuming.

As manufactured goods become less expensive in rela-

Figure 1.5 **CLINTON ADMINISTRATION**

Manufacturing jobs during the Clinton administration were very robust, peaking at 17.7 million in June 1998. Job numbers then began declining, dropping to just under 17 million when Clinton left office in January 2001.

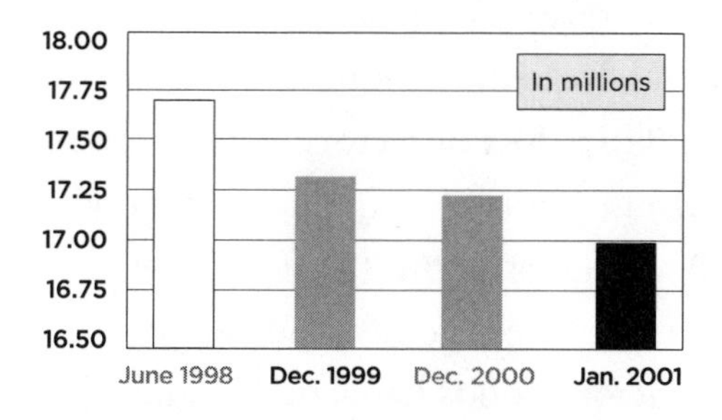

Figure 1.6 **BUSH ADMINISTRATION**

Manufacturing jobs during the Bush administration took a steep dive after Bush took office in January 2001, dropping to a low of 14.2 in January 2004. Since then jobs have increased slightly to 14.5 million.

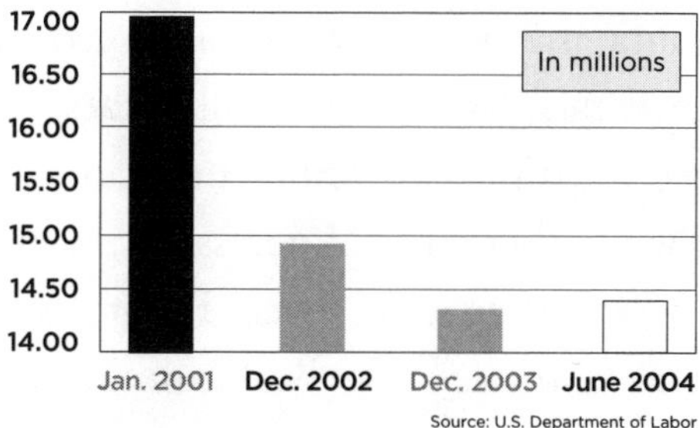

Source: U.S. Department of Labor

tion to services, Americans are consuming services at a more rapid pace than goods. As incomes rise, Americans are more likely to go out to dinner and see a movie, take a vacation, open a mutual fund, or get more healthcare than they are to buy a car or new furniture.

During the Clinton administration, slower growth in the consumption of manufactured goods was responsible for an estimated 17% reduction in the share of manufacturing jobs. Simply stated, had American consumption of manufacturing goods grown as fast as their consumption of services in the 1990s, manufacturing employment would have been 17% higher than it was in 2000.

However, during the Bush administration, the pace of job loss due to relatively slower growth in manufacturing demand has accelerated, accounting for approximately one-third of the relative loss of factory jobs. One reason is that while the consumption of manufactured goods went up in the Clinton administration (2%)—although less rapidly than nonmanufactured goods and services (14%)—consumption of factory goods has actually fallen 3% during the Bush administration.

This falloff has been particularly acute because of the bursting of the high-tech bubble. Simply, companies and consumers who purchased large quantities of high-tech goods in the late 1990s needed less of the same items in recent years.

For those of us covering the manufacturing sector, we are the first to admit that many of the published numbers are anything but accurate. What's more, all of these numbers can be manipulated to paint many different pictures. It's important to note the numbers are not necessarily black and white; there are many gray areas to consider. Regardless of what the numbers say, even though the Bush administration is not

Figure 1.7 **MANUFACTURING JOB DECLINE**

Beginning in August 2000 and ending January 2004, manufacturing lost jobs 38 out of 41 months. The only months that saw an increase were June 2002, June 2003, and August 2003.

August 2000 to December 2000

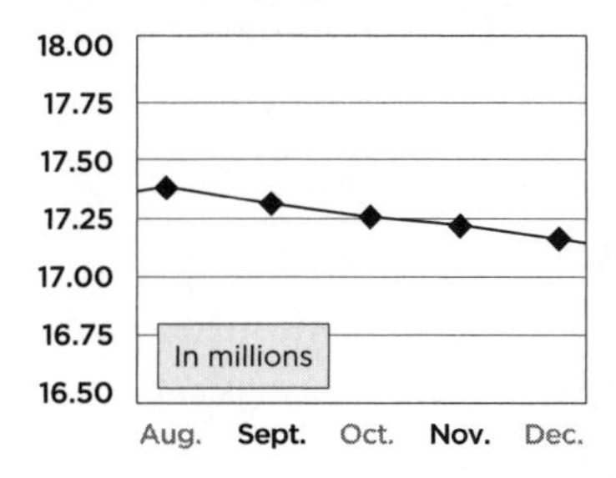

2001

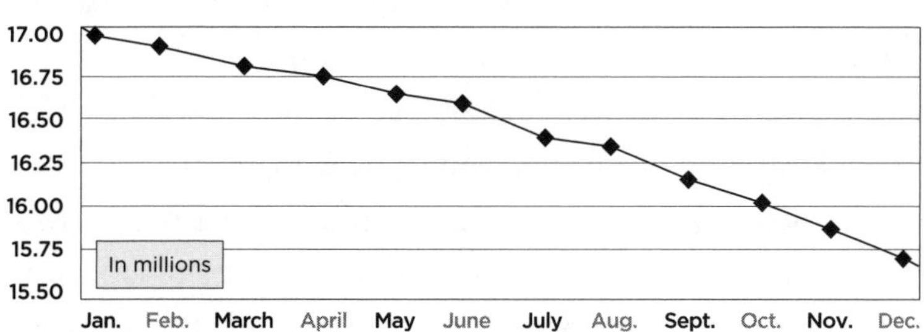

2002

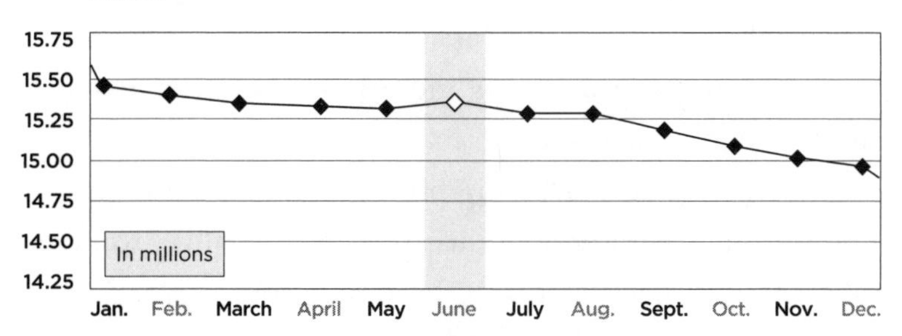

2003

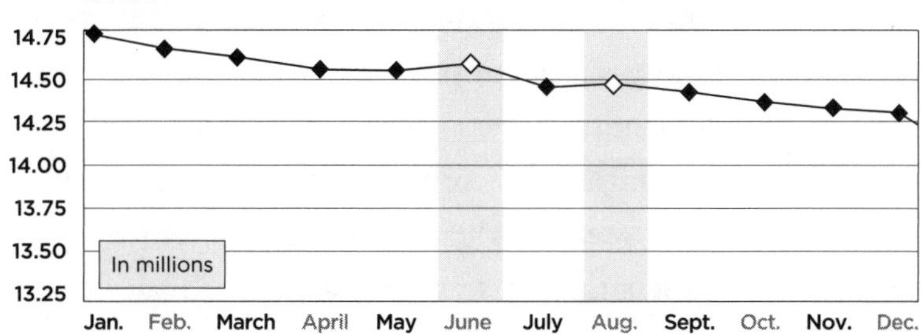

Source: U.S. Department of Labor

directly responsible for the dramatic loss of manufacturing jobs, it has done little, if anything, to reverse them.

First, it is a fact that during the past several decades the American workforce in manufacturing has been declining at a very steady and perhaps even a rapid clip compared with other nonfarm employment.

As of January 2004, manufacturing had lost jobs for 38 out of 41 months, the longest such stretch of monthly job losses since the Great Depression. The lone months that saw an increase were June 2002 with an increase of 64,000 jobs, June 2003 with an increase of 41,000 jobs, and August 2003 with an increase of 39,000 jobs. During the past 18 months, the United States has witnessed the weakest manufacturing recovery from a recession since the Federal Reserve started keeping tabs on such things back in 1919.[7]

Manufacturing employment has plummeted to its lowest level in more than 40 years—and it continues to fall with no end in sight. Therefore, it's time to strengthen American manufacturing and stem the tide of job losses that have gone largely unheeded.

No statistic captures this more than the sheer magnitude of the jobs that have permanently disappeared. From 1979 to 2001, manufacturing production jobs plummeted by more than 2 million. Since the end of 2001, some 2 million jobs in manufacturing have been lost.[8]

It's not just that jobs are fading away, but so many of the lost positions that help sustain our neighborhoods were good-paying, family-supporting careers. Far too many of today's manufacturing workers receive paychecks that just haven't kept pace with other professions. The purchasing power of manufacturing workers' earning per hour is currently about 7% less than it was 20 years ago. That's even taking into account workers putting in long overtime hours.

Digging deeper into the issue of lost jobs in manufacturing, during the past two years union jobs have been disappearing and the number of nonunion U.S. manufacturing jobs has in contrast actually increased by 19%. Last year, only 14.8% of U.S. manufacturing workers belonged to unions, compared with 35.4% in 1979.[9]

It must be reiterated again that most of the economic slowdown has nothing to do with the current presidency. Instead, there are three factors that are significantly impacting the market and they are: a big push by companies in the 1990s to purchase technology left little buying opportunities for the beginning of the new century; consumers started buying cheaper goods that China had flooded the market with; and productivity in general increased with advances in technology, leading to fewer workers needed. As technology has advanced, the need for workers has declined. In some ways many of the job losses in manufacturing are reminiscent of what happened to the agricultural sector during the Industrial Revolution. As expected when productivity increases faster than demand, employment decreases and employees are forced to move into other industries and fields. Productivity gains are clearly a contributing factor to what is happening in manufacturing. While manufacturing output remains fairly high, Americans have been able to produce more goods with less input and less workers overall. Most of the jobs lost will not be recovered.

In fact, some workers are moving into the services industry, which is said to be experiencing some increases where demand is picking up more rapidly than manufacturing productivity. In addition to productivity gains, some of the impacts on the manufacturing crisis can be attributable to the economy's most cyclical downturn and the deflating bubble of the 1990s. However, none of these events sufficiently explain the dire straits manufacturing is currently facing. Today's manufacturing crisis goes much deeper.

At the same time, the trade deficit in manufactured goods continues to climb to unprecedented heights, to more than $450 billion during the same period, or $1.2 billion each day. Today, the trade deficit has reached an unhealthy level of about $45 billion per month.[10]

China is playing a key role in the loss of jobs. What's more, as incomes begin to increase in that country it, too, will have to compete on the basis of education, not just cheap labor. The new wisdom on China is that its economy will implode. Some doubt whether this will happen. However, if China continues to grow at the rate it has been growing—some 11.5% in 2003—it's probably questionable that it will implode, however, it is truly headed for a crisis.

It's not surprising that the United States has a huge trade deficit with China, but what is new is that the majority of goods manufactured there is more than just apparel and furniture items.

There has been an increasing trade deficit when it comes to high-tech products and the trade deficit on items such as computers, electronics, communications equipment, and consumer audio and video equipment. This indicates a shift in the focus of this type of manufacturing activity away from the United States, affecting sales, of course, but also the foundation and potential growth of the industry.[11]

In the short run, some U.S. manufacturers of these products are well positioned to take advantage of the shift toward low-cost overseas assembly operations in China and other countries. In the longer term, however, China's growing skill base and the tax advantages are likely to take away that advantage, a factor that has led American companies to call for policy changes in the United States.

Of the $17 billion increase in the U.S. trade gap of these types of products in 2002, close to half ($7.2 billion) was due to the change in trade flows with China alone. U.S. imports from China rose sharply, while exports to China rose slightly.[12]

President Bush has begun to look at the manufacturing crisis a bit more seriously and is desperately trying to show that he has a solid plan. The president announced in early 2004 his plans to appoint someone to pay even closer attention to manufacturing. This manufacturing czar, whose formal title is assistant secretary of commerce for manufacturing and services, will be responsible for keeping an eye on the proverbial bouncing manufacturing ball. To date, Bush's political adversaries insist the administration's actions have been nothing more than public relations gimmicks designed to appease manufacturers. Instead of creating a position that would help ensure the health of the manufacturing sector, the administration has softened its impact by including services in this person's portfolio. Rather than improving the focus on manufacturing, the government appears to be focusing on the overall economy. In its zeal, the administration doesn't want to show its support for any one vertical market sector. However the creation of an assistant secretary of commerce for manufacturing and services gives manufacturers hope. The hope was that perhaps someday Americans might see an entire department—like that of the Department of Agriculture.

For a full recovery to truly occur more people need to find work and manufacturing must improve considerably here and abroad.

U.S. Under Secretary for International Trade Grant Aldonas insists the administration is taking some steps and the powers-that-be in Washington have a plan to help American manufacturers compete better in this global economy.

A concern by many manufacturers when it comes to competition is what they call unfair trade practices by foreign countries, particularly China. Aldonas has stated many times that Washington has pushed for liberalization of foreign markets and will continue to do so. Regarding China, Aldonas insists Washington, through its newly formed office of unfair trade investigation, will put more pressure on China to help manufacturers investigate unfair trade practices.

Some of the concerns of today's manufacturers that have yet to be answered by the government are:

- **What can be done for certain industries such as machine tool, tool and die, and textiles that are having unusually high penetration by foreign imports?**

- **How can small and midmarket manufacturers (SMM) compete with foreign companies that are state-owned enterprises?**

- **Will tax incentives be provided for companies making products in the U.S.?**

- **What can be done about intellectual property theft by foreign companies?**

It appears the Department of Commerce is gaining an appreciation for how critical manufacturing is to the survival of the U.S. economy. Without question the United States is the world's leading producer of manufactured goods. What's even more impressive is that the administration is starting to focus more attention and resources to help rejuvenate the industrial community. A prosperous and successful industrial sector is critical to creating better jobs, fostering innovation, increasing domestic productivity, and creating higher standards of living for all Americans.

As part of its plan to revitalize American manufacturing the Department of Commerce has come up with a series of recommendations designed to address the challenges faced by U.S. manufacturers.

Evans insisted that these recommendations represent the start of a process, and the administration's commitment to begin mending manufacturing. The recommendations are:

- **Enhancing government's focus on manufacturing competitiveness.**
- **Creating the conditions for economic growth and manufacturing investment.**
- **Lowering the cost of manufacturing in the U.S.**
- **Investing in innovation.**
- **Strengthening education, retraining, and economic diversification.**
- **Promoting open markets and a level playing field.**

While these recommendations are a step in the right

direction, they fall short of helping small and medium-size manufacturers today. With its meekest of actions, the government has yet to outline a comprehensive plan that can be acted upon today, rather than tomorrow.

With each passing day, it becomes painfully apparent that Democrats and Republicans are only looking at the rapid decline in manufacturing and the loss of American jobs as a political problem to be finessed in this election year, rather than a major economic crisis that can dramatically hinder the economic growth of the United States.

The fact that the Bush administration does not want to link trade agreements with the manufacturing crisis only reiterates the lack of true understanding of the trouble facing the United States. Since it is an election year, the presidential candidates have to appear to have an aversion to recognizing the bad news, are not willing to upset the so-called "free-trade" agreements that are currently in place.

And if it is true that some 3 million factory jobs have been lost since the mid-2000 peak, it is very realistic then to assume that U.S. manufacturing will never recover the ground lost to overseas competitors. The prolonged wage slump triggered by the overseas migration of some of America's best-paying jobs has been rippling through the U.S. economy and American society for at least two decades. The loss of these important jobs represents a shrinking of the employment base needed for a middle-class standard of living, stable families, and the reduction of local and state tax revenues.

In addition, the manufacturing crisis raises serious questions about the U.S. economy's ability to maintain a high-tech world-leading military without worrisome dependence on foreign products and technologies. Although it is true that defense-related imports come overwhelmingly from long-time allies or traditionally friendly countries, it is just as true that they are growing rapidly at a time when major disagreements increasingly mark the relationships between the United States and these countries.

Further, the massive loss of tax revenue, both corporate and personal, directly attributable to a disappearing industrial base, will undoubtedly constrain America's ability to sustain military operations in both peacetime and wartime at levels that U.S. policymakers have come to take for granted. Thus, the country faces a future in which the ability to project power and thereby affect events and outcomes over the world will be much more limited than at any time in the last century and a quarter.

Even more worrisome, the decline of American manufacturing is quickly feeding on itself and appears to be gaining momentum as overseas manufacturers continue to invest in the latest R&D and technology. Unless American manufacturers work together to help themselves, American producers could be permanently crippled.

Washington's continued failure to secure equitable "fair trade" status to compensate for the outsourcing is only increasing the plight of American manufacturing, and its rippling effects will be felt by subsequent generations.

Unless the government acts expeditiously, U.S.-based firms will continue to export jobs that Americans need and will continue to force more plant closings throughout the United States. Economists' assurances about the broad efficiencies and benefits of global commerce can ring hollow to someone who has spent his life building and fine-tuning his company only to watch its prospects disappear right before his eyes.

2

Economic Recovery

As of this writing, all economic indicators pointed to a recovering economy, and also to a manufacturing recovery. While the United States might be experiencing a recovery, the manufacturing recovery is not as strong or as long as the economic recovery. Unemployment overall has improved slowly, but even slower in manufacturing.

While employment is moving upward, albeit ever so slowly, these facts are not at all clear to the public. The bevy of Democratic presidential candidates in early 2004 had been emphasizing the worst prospect for the economy and unemployment.

Let's look at the record.

The highly respected manufacturing report by the Institute for Supply Management (ISM), which is a bellwether for future evaluations of manufacturing activity, says the PMI (originally termed the purchasing-manager index) was

61.1 in June 2004, a decrease of 1.7 points from May but still growing for the 13th consecutive month. The PMI is a composite index based on the seasonally adjusted diffusion

Figure 2.1 **PMI INDEX**

	Jan.	Feb.	March	April	May	June	July	Aug.	Sept.	Oct.	Nov.	Dec.
2004	63.6	61.4	62.5	62.4	62.8	61.1						
2003	53.0	49.4	46.6	46.2	50.0	50.4	52.6	55.0	54.7	57.1	61.3	63.4
2002	49.1	52.9	55.3	54.1	55.3	55.7	51.4	50.5	51.4	49.7	49.6	53.3
2001	41.3	41.1	42.8	43.1	42.0	43.9	44.6	48.1	47.4	40.3	45.1	47.3
2000	56.7	55.8	54.9	54.7	53.2	51.4	52.5	49.9	49.7	48.7	48.5	43.9
1999	50.6	51.7	52.4	52.3	54.3	55.8	53.6	54.8	57.0	57.2	58.1	57.8
1998	53.8	52.9	52.9	52.2	50.9	48.9	49.2	49.3	48.7	48.7	48.2	46.8
1997	53.8	53.1	53.8	53.7	56.1	54.9	57.7	56.3	53.9	56.4	55.7	54.5
1996	45.5	45.9	46.9	49.3	49.1	53.6	49.7	51.6	51.1	50.5	53.0	55.2
1995	57.4	55.1	52.1	51.5	46.7	45.9	50.7	47.1	48.1	46.7	45.9	46.2
1994	56.0	56.5	56.9	57.4	58.2	58.8	58.5	58.0	59.0	59.4	59.2	56.1

Source: Institute for Supply Management

indices for five of the indicators with varying weights: New orders, 30%; production, 25%; employment, 20%; supplier deliveries, 15%; and inventories, 10%. The overall U.S. economy grew for the 32nd consecutive month.

Despite the doom and gloom reports from the news media, the economy grew 3.1% in 2003 and the White House forecasts the economy will climb 4.4% by the end of 2004, the most since 1999.

Consumption growth was about the same as in the fourth quarter of 2003, and with the consumer accounting for two-thirds or more of gross domestic product, consumption increase is an indication that growth is not slowing down, and may be even potentially picking up momentum. As the third quarter was coming to a close, excitement was stimulated when the GDP soared to 8.2%.

According to the government this is the largest increase in almost 20 years. In addition, U.S. factory activity is said to be surging at a remarkable pace. And to top that, some industry observers report that manufacturing activity is surpassing expectations and is growing at its fastest pace in almost two decades.

ISM comments from purchasing and supply managers have become increasingly optimistic as more and more industries indicate improvement. The general tone of the panel has improved significantly the first part of 2004. Its only major concerns appear to be steel availability and energy prices.

Thus, on the economic front, key indicators showed the economy is recovering, which is in stark contrast from just a little more than a year ago when manufacturing was petering out. In fact, the March 2003 Manufacturing Report on Business by the ISM showed the economy failed to grow that month, marking the end of four consecutive months of growth.

But more than a year later manufacturing had made a strong comeback as the June 2004 ISM Manufacturing Report on Business showed manufacturing was recovering nicely and all indications pointed to even more growth regardless of who wins the 2004 presidential election.

ISM's New Orders Index grew in June 2004 with a reading of 60. The index is 2.8 percentage points lower than the 62.8 registered in May, and it is the 14th consecutive month the index has exceeded 50. Anything more than 50 means manufacturing is growing rather than contracting.

Even the all-crucial Employment Index for June 2004

was at a robust 57.4, more than 10 percentage points higher than the March 2003 number of 46.2.[1]

Figure 2.2 **PMI INDEX 2004**

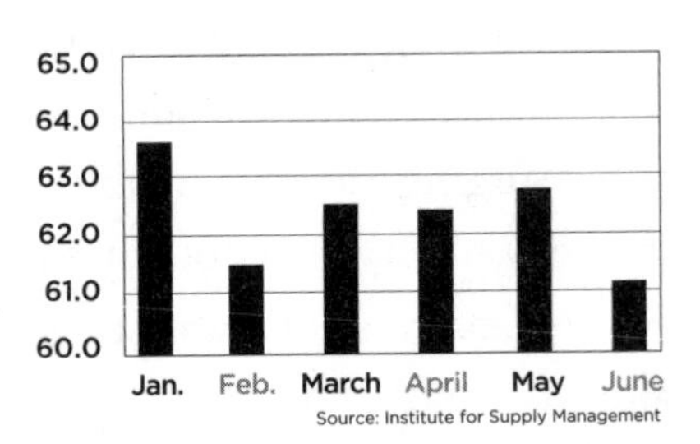

For the first six months of 2004, the PMI has been above the 60 mark—a stretch not seen since the latter part of 1983. The last time the PMI was at or even above 60 was in 1987, which is pretty incredible considering the white-hot days of the mid and late 1990s.

What's more, there are projections that manufacturing production growth would top 5.6% when all the numbers are in for first quarter 2004—exceeding the growth rate of the economy at large.[2]

The signs that manufacturing is getting stronger are beginning to add up and industry experts say the U.S. economy is on course for another solid quarter of growth.

Even the chief economist at the Conference Board says the GDP jumped slightly more than 4% in the first quarter of 2004.[3]

The very helpful leading index reported monthly by the Conference Board, auguring the future, increased 0.5% in January 2004, the largest increase since October 2003. The leading index has now increased at a 5% annual rate from the most recent low in March 2003, and the growth has continued to be widespread, with five of the 10 indicators that make up the index increasing.

The positive contributors were the indexes of consumer expectations, stock prices, average weekly manufacturing hours, vendor performance, and average weekly initial claims for unemployment insurance. The negative contributors, worst to better, were building permits, interest-rate spread, real money supply, and manufacturers' new orders for non-defense capital goods. Remaining unchanged were manufacturers' new orders for consumer goods and materials.

While it is particularly heartening to see that the employment index jumped up well into positive growth territory, it is still far below where the industry needs it to be to continue to foster any significant long-term growth. With that said, however, with jobs coming back and the pickup in the industrial sector causing slower deliveries and rising commodity prices, these are all signs that manufacturing is on the mend.

ISM's backlog-of-orders index decreased in June 2004 to 58.5 from 63 in May 2004 following an increase in January, and the employment index was lower at 59.7 from 61.9 the month before. The price index indicates that manufacturers experienced higher prices for their purchases, but at a slower rate.

Since 1997, the productivity in U.S. factories increased to an annual rate of 5.1%. According to the number crunchers that is the fastest sustained gain in at least 40 years, and well ahead of the heyday of the 1960s. Ironically, despite these gains, manufacturing output is still down about 4% from its 2000 peak of 139.1. Not surprisingly, imported goods are up 8% between 2002

Figure 2.3 **MANUFACTURING OUTPUT**

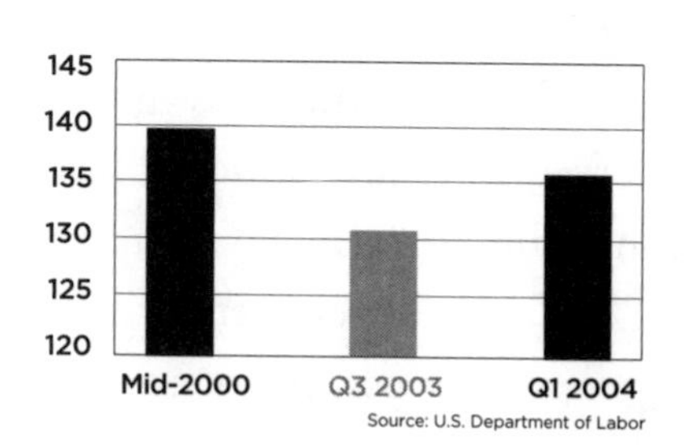

and 2003.[4]

Since mid-2000, U.S. manufacturing output has declined from the high of 139.1 in mid-2000 to just over 130 by the second quarter of 2003. Output, though, is beginning to climb as the latest numbers available, first quarter 2004, show an increase to 135.4.[5]

Greater productivity means stronger economies and more competitive manufacturing. The United States is believed to be a world leader in productivity, and the European Union (EU) has been flat. However, the 10 mostly Eastern European countries joining the EU in May 2004 have overall greater productivity than the United States, which means the EU as a whole could come close to matching the United States.

The countries are Czech Republic, Estonia, Cyprus, Latvia, Lithuania, Hungary, Malta, Poland, Slovenia, and Slovakia. Part of the reason for their recent productivity growth is that many of their economies were stultified in the past by Soviet control and were at a lower base from which to grow.

Productivity is a measure of labor productivity, the quantity of labor required to produce a unit of a product. In economics, productivity is measured as a country's GDP per capita of the employed population.

Productivity growth depends on the quality of plant and machinery (physical capital), improvements in the skills of the labor force, technological advances, and new ways of organizing. Productivity growth currently is significantly determined by investment in information and communication technology (ICT).

ICT is a core element of knowledge in the economy and an important complement to R&D. Those EU members recording rising productivity levels close to those of the United States are the ones in which the use of ICT is widespread. Productivity gains are closely related to the use and diffusion of ICT. EU's ICT spending is weak overall.

The respected Conference Board, which has more than 2,000 company members in 66 nations, reports that EU productivity growth is 0.8%, lagging far behind the United States at 2.6%. The 10 new members, though, had an impressive average of 4.2% during the past eight years.[6]

The new members will significantly help the EU fulfill its goal of becoming by 2010 "the most competitive and dynamic knowledge-based economy in the world, capable of sustainable economic growth with more and better jobs and greater social cohesion." This objective was adopted by the Lisbon European Council in 2000.

While the U.S. has traditionally outpaced Europe in productivity growth by profitably exploiting new information and communication technology, change may be on the horizon if action is not taken. This is especially true in the trade and financial sectors.

Although real investment in information and communication technologies in Europe has grown about as fast as in the U.S., Europe is beginning from a much smaller base. The lack of a fully integrated EU market also makes it more difficult to take advantage of the new market opportunities. However, Europe is making progress at what can be considered a very aggressive pace. The Conference Board says despite the projected productivity growth gains for the European Union, catching up with the United States is a strong challenge. While the EU might not be providing the sustained solid growth needed to lead global prosperity just yet, its prospects continue to become even more optimistic than in previous years.

The Board forecasts that U.S. economic growth will reach 5.9% in 2004, with global growth at about 5%. While growth is healthy in both Europe and Japan, their growth rates will remain in the 3% range.

Interestingly some organizations have been also reporting that not only is American manufacturing activity beginning to show promise of a solid recovery, but strong performances are being experienced throughout the rest of the world.

Global manufacturing economic growth continued in June 2004, building on the positive performance recorded throughout the second half of 2003 and in the start of 2004.[7]

This assessment was based on the Global Manufacturing PMI of 55.9, a composite index produced in association with the ISM and the Intl. Federation of Purchasing and Materials Management (IFPMM), Aarau, Switzerland, an organization of 42 national associations and about 200,000 procurement professionals. Much like the fluctuations of the U.S., the PMI was down from 56.9 in January, then up again in May at 57.6, then down again in June at 56.5. This indicates the strength of the global economy simply rests on the productivity and growth of the United States—a fact that continues to be ignored every day by the media and the U.S. government.[8]

Figure 2.4 **GLOBAL MANUFACTURING PMI**

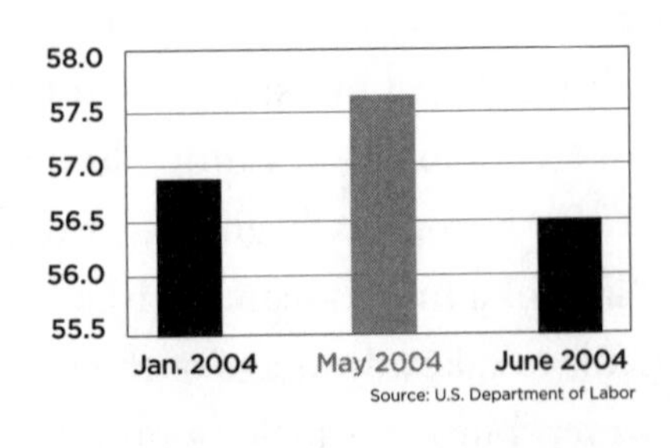

The Global Manufacturing PMI is based on responses

from approximately 7,000 companies worldwide operating in countries that account for roughly 76% of total global manufacturing output. Information provided by each national association relates only to the domestic production facilities in each of the countries.[9]

PMIs in all except one of the 19 countries for which February data were available pointed to expansion. However, the rate of expansion indicated by the PMI slackened during the month, reflecting slower growth in 11 of these countries. Expansion of the global manufacturing economy was led by the United States, which recorded the sharpest growth of all nations for the fourth month in a row.

Market expansion was also reported in Japan and to a lesser extent in the Eurozone and the United Kingdom.

The Global Manufacturing Output Index of 58.1 in February indicated expansion of manufacturing output for the 10th month running and at its present level is consistent with annualized growth of worldwide industrial production of about 9%. June numbers were statistically unchanged.

However, after falling from 60.9 in January, the index suggested slower growth than in the previous month. The United States, Japan, the Eurozone, and the United Kingdom all recorded further robust increases of production, although the rates of growth in each of these economies eased from a month before.

The Global Manufacturing Input Price Index rose to 66.5 in February from 61.8 in January, and then shot up to 72.1 in June, continuing its upward trend and indicating further sharp inflation of worldwide material prices.

The Global Manufacturing New Orders Index registered 59.2 in February, indicating further substantial growth of new business, although the rate of increase eased for the second month in a row and continued to fall in June to 57.1. All of the major industrial nations—United States, Japan, Eurozone, and United Kingdom—recorded marked growth of new orders, although at slower rates than a month earlier, with the sharpest rise reported by the United States.

The Global Manufacturing Employment Index increased in February for the third successive month to 51.1, considered only a marginal increase, and was up some more at 53.2 in June, although the United States reported marked growth. In contrast, employment declined in Japan and Eurozone, with the four largest Euro-area economies reporting lower employment.[10]

There is no doubt that a rising trade deficit in goods, expected to surpass $500 billion by the end of 2004, has played a pivotal role in the weakening economy.

According to reports, in 1990 almost 60% of U.S. imports from China consisted of apparel, furniture, rubber and plastic products, leather and leather products, and miscellaneous products. Higher ticket items or goods, such as nonelectrical machinery, electronics, and transportation equipment, accounted for only 16.9% of U.S. imports from China. By 2001, U.S. imports had dropped to 45.1%, while the higher valued items more than doubled to a whopping 35.6%. The big looming question is whether the government can turn the tide anytime soon.[11]

3

Reversing the Trend

The job rejuvenation and economic-growth tax relief bill that was passed in 2003 is worthy of manufacturers' attention because of the enhanced incentives it offers for new-equipment orders. Now is the perfect time for manufacturers to take a good hard look and perhaps take advantage of the tax cuts that the administration has instituted.

The legislation contains a new 60% expensing allowance for machine tools and other equipment ordered between May 6, 2003, and Dec. 31, 2004, and placed in service by Dec. 31, 2004. This replaces the temporary 30% expensing allowance enacted in 2002.[1]

Here's how the new provision works:

If a manufacturer orders a new machine tool costing $100,000, the firm can write off 57% of the asset in the first year and 69% over two years (compared with 14% and 39% under the old law). This adds up to a first-year tax cut of $15,050 on a $100,000 machine.

There's a special rule for small businesses (those whose equipment purchases of all kinds do not exceed $400,000). Small business gets to expense the first $100,000 until Dec. 31, 2005. The 50% expensing allowance can be taken on the remaining basis of the machine.

In other words, a qualifying small business that buys a $100,000 machine can expense it all in the first year. A $200,000 machine could qualify for $157,000 (conditions below) first-year deduction, 78.5% of the assets, and a $300,000 machine could quality for a $214,000 (conditions below) first-year deduction, 71.3% of the asset.

The conditions on the above assume a customer is in a seven-year asset depreciation class. For customers in the five-year class, the first-year tax saving is $21,000 on a $100,000 machine and the tax cut is $14,000.

While this is a step in the right direction, the government needs to work closer with manufacturers. Regardless of this incentive, even before we discuss making manufacturing a priority once again, government leaders need to get a better understanding of what is at stake if American manufacturing continues to decline. Additionally, American citizens need to really understand that cheaper prices on manufactured goods might be good in the short term, but could, and will have disastrous long-term results. What this means for all of us is that we need to force Congress to pay more attention to manufacturing and to take a longer more in-depth look at its problems today and how they will impact us tomorrow. It cannot be stated enough that the economic stability of future generations is at stake if all Americans do not act accordingly.

The key to reversing American manufacturing's decline

is to open up key foreign markets that still remain closed to U.S. producers.

Manufacturers do not need more trade agreements like many of those in existence today. Rather, they need trade agreements that open up markets that are currently closed to free and fair trade agreements. Too many foreign markets still remain closed to U.S. exports. The handful of economies wealthy enough to consume American-made goods have elected to put up more roadblocks hindering U.S. companies from competing across the global markets.

First and foremost, it needs to be emphasized that the fundamental cause of the manufacturing crisis is the cumulative and continuing impact of more than twenty years of misdirected U.S. trade and globalization policies. Washington has consistently failed to open foreign markets adequately to U.S. manufacturers—despite years of promises and fanfare that breed each new trade agreement. The federal government has come up short in its attempt to combat predatory foreign trade practices aimed at undermining the U.S. within its home turf.

What's even more disturbing is that the current administration and the one that came before it has encouraged manufacturers to open production plants in Mexico, China, and other countries.

As a result, this has triggered an overseas migration of America's best-paying, highly qualified jobs. Therefore, the massive loss of tax revenue, both personal and corporate, is directly attributable to an evaporating manufacturing sector that continues to have devastating consequences.

Most, if not all, of the trade policies to date have proven to be failures. Most are just wishful thinking with no solid foundation to make them really work.

Some have even proposed that we should trade only with countries that have the same level of health, safety, and environmental regulations. The problem with this view is that it would limit our trading partners to only a handful of rich nations, and dramatically raise the prices of goods and services for consumers, not to mention stifle industries that rely on imports.

The American government has also failed miserably to combat predatory foreign trade practices aimed at undermining U.S. producers in their home market. Perversely, Washington has responded to these failures by encouraging U.S. manufacturers to supply their home market from low-cost Third World production platforms like Mexico and China. And most U.S. multinational corporations, and indeed some of their smaller suppliers, have responded with enthusiasm. It's obvious that the government has done a lousy job of opening foreign markets to U.S. products.

It's important to remember as many people are beating the protectionist drum, the opportunities offered by global commerce—lower prices, more spending power for consumers, enhanced opportunities for exports—are forcing many smaller manufacturers to close their doors. While some companies are benefiting, others are suffering.

Pointing to imports, cheap foreign labor, or Chinese currency manipulation as the root cause of the problem is not completely accurate. Rather, they are simply symptoms of a much bigger illness. The root cause of much of the manufacturing crisis dates back to the North American Free Trade Agreement (NAFTA)-style trade agreements of the past decade that essentially encourage the movement of jobs and production plants overseas. The NAFTA agreements that were signed some 10 years ago have proven to be nothing more than to give companies a blessing when sending manufacturing capacity and jobs overseas. While changing into a service-based industry seems okay for many politicians, it will soon prove to be the linchpin in what brings America to its knees.

More jobs and manufacturing capacity would stay in the United States if domestic producers had more opportunities to sell their products abroad. In fact, U.S. producers would be generating more revenue, which in turn would lead to greater investments in people, jobs, and technology.

However, it appears the U.S. trade policymakers have focused more of their deal making on low-income countries too poor to purchase American products, but do have the ability to export products to supply the U.S. market. The U.S. government has rarely insisted on reciprocity in trade agreements, and has failed to monitor and enforce most of the agreements they have signed.

Decades of mounting trade imbalances demonstrate that neither Democrats nor Republicans have been effective in drafting trade agreements that can improve the nation's trade balance.

The laws and policies that have been drafted have proven futile in expanding the nation's production capabilities or its ability to combat the predatory economic practices of foreign firms that harm American manufacturers.

As a result, the manufacturers that remain in the United States find themselves handicapped in their efforts to sell abroad and to compete fairly in an open market, let alone reap the efficiencies these other countries are experiencing. It is a highly known fact that 1997 was the year that Asia, Russia, and Latin America suffered huge financial crises. The result now has been smaller export markets for American producers and more aggressive efforts by distressed foreign producers to dump their goods on the U.S. market. Thus, American consumption shows that much of the U.S. production capacity is gone for good. Consequently, the trade deficit will not return easily, causing greater concern about future growth.

China's strategic policies in this market call for the United States to identify and implement economically sound strategies to maintain the base of skills, knowledge, and physical capital underlying high-tech innovation.

According to a report released in October 2003, a massive expansion of semiconductor manufacturing capacity is currently underway in China. "China's Emerging Semiconductor Industry" report was prepared by Dewey Ballantine LLC, for the Semiconductor Industry Association (SIA). At present, by conservative estimates, 19 new wafer-fabrication facilities are operational, under construction, or planned in mainland China under Taiwanese majority ownership and management.[3]

Although these plants will not initially utilize state-of-the-art process technology, their managers plan to capitalize on technology transfers from multiple foreign partners to approach technological near-parity with the global leaders. As anticipated these new facilities are in response to replicating Taiwan's success in building a leading-edge semiconductor industry on a larger scale on the mainland. China is implementing promotional policies which succeeded in Taiwan and is drawing heavily on Taiwanese capital, technology, and managerial and engineering talent to build the new manufacturing facilities.

The single most important motivator for this movement is a measure implemented by the Chinese government in

2000 which creates a major cost advantage for mainland-based semiconductor production, says the study. All imported integrated circuits (ICs) are subject to a 17% value-added tax (VAT), while ICs designed or manufactured domestically are rebated everything over 3%, in effect a 14% difference.

China, currently the world's fastest growing major market for semiconductors, ranks as the third largest market in the world with $19 billion in sales, and it imports about 85% of its total consumption of semiconductor devices and nearly all of its advanced semiconductor manufacturing equipment, with U.S. firms accounting for a substantial proportion of total imports.[4]

While there are major concerns over China's growing semiconductor industry, some believe this market provides major export opportunities for the U.S. However, others dispute these claims—insisting that in the longer term, it could erode the U.S. microelectronics infrastructure and contribute to an eventual loss of U.S. leadership in the field, only reinforcing concerns over China's impact on U.S. manufacturing

SIA strongly opposes China's VAT system, saying the World Trade Organization, which China recently joined, does not allow countries to engage in activities that treat domestic producers and products more favorably than imported products.

While it's obvious we cannot control the insidious actions of other foreign countries, the American government does have the power and the authority to remedy the problems that exist here within the borders of the United States. Therefore, the crucial component to reversing this downward spiral in manufacturing lies within our government's managing of its own behavior and controlling access to U.S. markets. It probably goes without saying that promoting economic growth abroad is only a beginning. Finding a cure for what ails manufacturing will require a greater infusion of actionable ideas, rather than more rhetoric.

4

Why Is Manufacturing Important?

How important is manufacturing to the nation's economy? Some argue that the so-called real economy is about making widgets and other things, while providing services is a second-class activity that doesn't generate the necessary wages. In reality, the economic stability of the country is contingent upon the productivity of the entire economy, not one particular sector.

Turning to more of its strengths, manufacturing is essential to economic growth and employment opportunities within the United States. Manufacturing historically has been a major generator of good, high-skilled, well-paying jobs that, in turn, generate four additional jobs in the economy. Manufacturing's decline is undermining the quality of these jobs and contributing to the stagnation in workers' wages. The massive plant closings and job layoffs that have occurred during the past three years are contributing factors to the fiscal crises afflicting just about every state in the nation. What's more, the rising costs stemming from government regulations and other legislative requirements are proving to be too burdensome for many companies, often forcing them to choose between laying off plant workers or outsourcing to foreign countries such as China.

Manufacturing has been the primary driver of productivity gains, technological innovation, and economic growth. A robust domestic manufacturing base is vital for maintaining a strong defense and homeland security. The loss of manufacturing capacity will certainly weaken America's leadership in critical technological areas and limit long-term productivity growth. The dependence on foreign sources for strategically critical products and components is in itself a threat to the nation's defense.

Regardless of your opinion, the manufacturing arena provides many well-paying jobs, but most of the service sectors have some of the highest paying jobs in the United States, exceeding those in manufacturing. Despite the debate, manufacturing accounts for the largest component of the U.S. trade balance and a weak manufacturing sector leads to an even larger trade deficit. Concerns over the trade deficit growing will only intensify if more products continue to be imported leading to a significant deterioration of the value of the dollar. Ultimately, this will force Americans to pay more for imports. Alongside of our ability to compete, manufacturing plays a critical role in determining the trade balance. And that means producing products other countries need and which have the money to buy the goods the U.S. makes.

American manufacturing workers are the most productive in the world. But they operate under enormous competitive disadvantages: unfair trade and tax policies, an overvalued dollar and undervalued foreign currencies, inadequate investment incentives, healthcare costs not borne by overseas producers, lack of a level playing field between older firms and both foreign producers and newer firms with younger workforces and fewer retirees, and foreign government subsidies. These problems must be addressed immediately or American manufacturing capacity and jobs will be lost permanently.

The depth of the nation's economic recovery and economic prosperity is contingent upon whether we can successfully revive the manufacturing base in the United States. Without question this means expanding manufacturing exports and reducing the dangerously large deficit. Ultimately, the goal is to return the trade deficit to a positive balance. If this turnaround is not acted upon quickly, American manufacturing could falter to unprecedented lows that are not recoverable, which could result in a prolonged financial crisis never seen before.

Beyond looking at our economic security, it is also critical to examine manufacturing's contribution to our national security. This contribution is extremely visible to the men and women who serve in the U.S. military, but it is not as apparent to everyday citizens. The U.S. military understands that to effectively safeguard civilians while simultaneously targeting terrorists, it needs everything from advanced fighter planes to high-tech guided weapons systems, and digital and laser communications that comprise state-of-the-art command and control. None of this would be possible without the advances of U.S. technologies. Without question, manufacturing has proven to be the driving force of technological change and productivity growth unmatched by any other country.

UNDERSTANDING MANUFACTURING

Even before we can begin to solve the crisis in manufacturing we need to truly understand what makes up the small and medium-size manufacturer (SMM), or otherwise known as SME (small and medium enterprise). First and foremost, manufacturing companies in the U.S. are not a homogeneous group. While the government suggests that the definition of small manufacturer is any company with less than 500 employees, industry publication *Start* magazine classifies manufacturers in three sectors: Large manufacturers are those with more than $1 billion in revenue; midsize companies are those between $200 million and just under $1 billion; and small manufacturers are those with fewer than $200 million in annual revenue.[1]

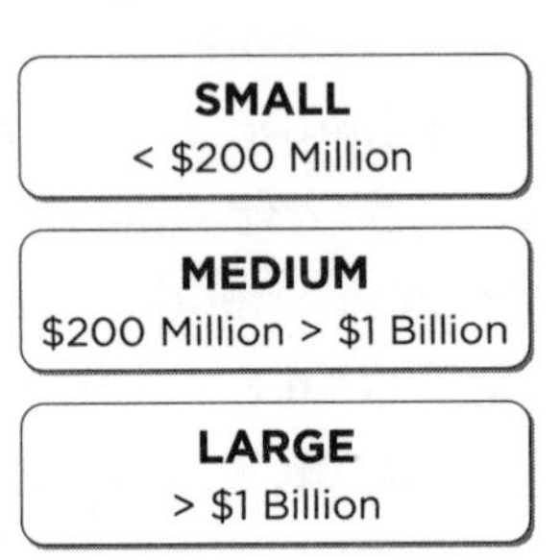

Mike Collins, a consultant with more than 40 years experience in manufacturing, accurately segments manufacturers into four categories: survival mode, family mode, professionally managed, and giant public. These manufacturers are different in that he shows there are logarithmic differences in the four types in terms of resources, knowledge, experience, staff, and the wherewithal to deal with change.

1. Classification of manufacturers

First is the fact that manufacturing companies in the U.S., as noted earlier, are not a homogeneous group. In fact, there are not simply large and small manufacturers as the government suggests in its definition of small manufacturers being any company of less than 500 employees. *Start* magazine's classification of the manufacturing sector more accurately reflects the market and how small and medium-size companies categorize themselves.

2. Resource limitations

The second important point is that all small and midsize manufacturers are restricted by resource limitations. "FACTS" is an acronym that accurately characterizes the reality of the small manufacturing environment described as:

F - Fear of making a wrong decision
A - Limited access to capital
C - Cashflow problems
T - Time constraints
S - Small or no staff

3. SMMs are not small versions of larger manufacturers.

Third, and perhaps most important, is the understanding that small manufacturers are not small versions of large manufacturers. One of the most popular approaches to helping SMMs or SMEs is the assumption that the same theoretical concepts or solutions pioneered by large, world-class household-name manufacturers will also work in SMMs. Most of these new solutions are acronyms such as MRP, TQM, ISO, ERP, MES, SCM, CRM, SRM, PLM, etc.; and there is usually a new solution for manufacturers every year. Even though the principles of the solution may be correct, to be useful to SMMs the solutions must be customized to fit the growing needs of small and medium-size firms and the level of development of the company's management systems. In a medical analogy, one must be careful that the dose of medicine does not kill the patient or make the symptoms worse.

To develop any kind of a plan or attempt to mend American manufacturing one must first identify who these manufacturers are and what are their special needs. And, the first question that should be asked is what can be done (or should be done) to help Type Four manufacturing companies. These are the publicly held, multinational companies that drive all supply chains. They are the companies that are closing U.S. plants, moving product lines to other countries, sourcing products and services from foreign suppliers, and relentlessly driving cost reduction in their supply chains. Should we all be concerned that China's currency is floating or that these companies consider their taxes, health costs, insurance costs, and environmental costs too high in the U.S.? These giant manufacturers have the financial wherewithal to fund many lobbyists working on their behalf and they clearly have the ear of the White House and the Congress.

Be it right or wrong, manufacturers are most likely going to do whatever they see is most profitable for their shareholders, and there is probably nothing that their suppliers or anyone else can do about it. They will probably all survive in this new era of globalization, even if they don't manufacture products in the U.S. It cannot be stated enough that what Type Four manufacturers are doing is short-sighted and will hurt American manufacturing and the country as a whole in the long run. It would be helpful if Type Fours could take a longer view and work at developing true alliances with their suppliers and maintaining the critical mass of U.S. manufacturing for the good of the entire economy. It's apparent that all Americans should be concerned about mending American manufacturing and should not assume Type Four companies are going to change unless there are financial advantages to change. Americans simply cannot depend on their altruism, patriotism, or loyalty to the U.S. manufacturing base.

So, the real problem boils down to how to help Type One, Two, and Three manufacturing companies find ways to compete in this number-crunching, high-return world to which we live. If Americans assume that sourcing offshore will continue along with relentless cost reduction, then it is also safe to assume that all customers will continue to change, will leave American suppliers, or force suppliers to abandon them. This means that SMMs or SMEs are faced with doing business in a changing environment with moving targets.

5

Globalization and Innovation

Innovation has enabled the optimization of manufacturing plants for decades. Innovation can and has enabled some U.S. manufacturers to boost productivity and at the same time expand their output despite the increased pressures from foreign competition.

So let's evaluate then the real causes for the manufacturing crisis. Some blame competition from low-wage offshore factories, an excessively strong U.S. dollar, high corporate taxes, and the rising bill for employee and retiree benefits, and others blame outsourcing and a lack of free and fair trade. In reality, the real problems facing American manufacturing are a combination of many factors. However, one of the biggest crippling factors is a lack of research and development and innovation. That's technological innovation. While it's easy to point a finger at many of the outside forces that continue to impact the instability of U.S.

manufacturers, we need first to look within our own backyard.

Throughout the 1970s and 1980s discussions centered around the globalization of industry. However, it wasn't until the 1990s that the true impact of globalization was actually realized. Going global has certainly been a tricky business riddled with intense competition and unparalleled pressure. Achieving a successful globalization strategy required businesses to be aggressive, prepared, and visionary. Historians have coined the 1990s as the decade of globalization, noting that it was an era of acquire or be acquired. The fiscally strong companies gobbled up the weaker ones only to increase the overall size and dominance in the markets they served. Ironically, some of the largest companies in the world, which lacked the financial wherewithal to withstand a hostile takeover, became the victims of an ever-increasing acquisitions race.

The emergence of a global economy has brought unbelievable access to goods and services and with it, unrelenting change.

Today, globalization has proved to be very challenging for many companies, which ultimately has led to decreases in workforces and increased productivity. Therefore, much of the downsizing was the result of technological advances which help to reduce labor without hindering productivity.

Several other trends have come as a result of globalization as well. Among them are new pressures from other regions of the world and greater challenges in accessing raw materials. While some manufacturers were able to focus on one geographic market segment, new competitors from emerging industrial regions are entering these markets with a flurry of activity. Now American manufacturers face more competition from global competitors. Raw materials are easier to access and less costly giving global competitors a favorable advantage.

As a result, manufacturers in the industrialized regions were forced to pay higher wages for the higher skills and better education of their local workforce. What's not surprising though is that the education and skill gap is narrowing presenting manufacturers in developed regions with a highly competitive marketplace.

Now manufacturers need to adapt to their ever-changing market conditions. There is a new paradigm shift taking place in manufacturing. Globalization is certainly creating new challenges and opportunities for U.S.-based firms. In the past, manufacturers could leverage marketing strategies and superior sales to speed past the competition. However, moving forward, manufacturers are being forced to win deals through advanced manufacturing strategies and a better execution and delivery of these efforts.

Manufacturers can meet these new challenges head-on with the right technological investment that ultimately increases plant performance and efficiency. Technology does offer tremendous opportunities to improve the manufacturing process. While all this appears to make sense, many manufacturers have failed to keep pace with research and development to spark the necessary technological change.

Figure 5.1 **FEDERAL RESEARCH AND DEVELOPMENT SPENDING**

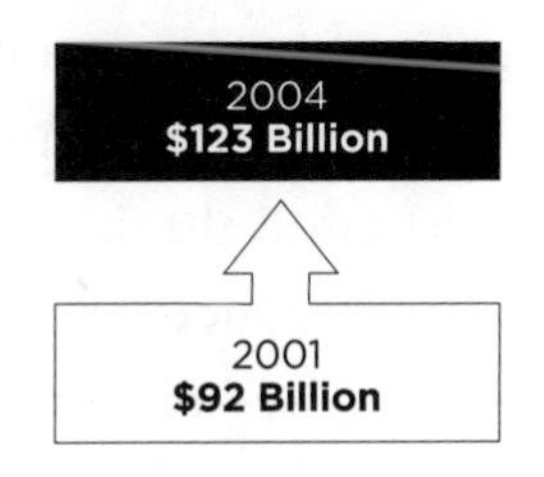

In an effort to rejuvenate investments, President Bush has provided a renewed focus on federal research and development by proposing a record $123 billion budget in fiscal 2004, an increase of more than 34% over funding levels that existed when he took office.

Some reports say that between 1995-2000, with the exception of high tech, pharmaceutical, medical equipment, and automotive, many manufacturers scaled back their investments in domestic research and development. Therefore, investment in technology such as software and capital equipment is not as high as it should be to continue to make American manufacturers powerhouses. However, a report by the National Association of Manufacturers, says manufacturers doled out $127 billion on R&D in 2002, which accounted for 67% of the total R&D dollars spent by the private sector. This percentage has been fairly constant the past 20 years, as approximately 60% of total R&D spending in the United States has come from manufacturing firms.[1]

With some key R&D initiatives, there are basically three key areas where government can help:

First and foremost, the U.S. government can help companies compete by encouraging even more innovation. To elicit such high performance, the government needs to draw out the passion, enthusiasm, and drive that exists within the manufacturing sector.

Now is the time for manufacturers to take a good long look at how they can make their plants more efficient and effective, reduce errors, improve their assets, and ultimately make their plants more profitable. What that entails is driving innovation through developing the next generation of innovative products and by using leading-edge technology to achieve those key initiatives.

Creating the right R&D initiative will encourage greater investment in essential programs. Some of the existing programs, such as the Manufacturing Extension Partnership (MEP) and other grants, will encourage manufacturers to improve their technology and equipment. MEP plays an important role in securing the future of the nation's small and medium-size industrial base.

Once a greater investment is made, U.S. manufacturers will become more competitive, which in turn will restore manufacturing demand as it leads to a rebound in the capital-goods sector. The goal is not to boost demand for manufactured goods, but rather, increase the competitiveness of U.S. manufacturers. With the right policies by the current administration, this will not only lead to increased exports, but also reduced imports, particularly in high-value-added manufactured goods.

Ultimately these moves will reinvigorate manufacturing and increase employment across all sectors. Therefore, technology will play a critical role in any manufacturing policy that has teeth. Some of the most promising areas include machine-to-machine communications, wireless, and the use of new materials such as nanotechnology and intelligent process-control systems.

The administration can play a pivotal role by becoming partners with manufacturers for the widespread deployment of these technologies and many others. Just as the federal government played a critical role in past manufacturing innovations—including computer-controlled machines, optics, robotics, and composite materials—it can drive the next manufacturing transformation.

In order to jump-start manufacturing, Congress needs to help SMMs gain greater access to foreign markets. While the opening of new markets to exports through trade agreements is essential, the government should focus just as much attention on helping manufacturers access those same markets.

Manufacturers need to rally government officials to set up training and education programs that provide the necessary information to establish global enterprises. The government is familiar with the dealings of foreign nations and can share this experience with small and medium-sized manufacturers that do not have the resources to do it on their own. A more tightly linked global initiative can make all the difference in helping U.S. manufacturers gain access to international markets.

Manufacturing also plays a critical role in protecting the nation. The making of advanced fighter planes and high-tech guided-weapons systems, and advanced digital and laser communications that comprise state-of-the-art command and control are proven products from innovation. These technological innovations must be protected and Washington needs to do everything in its power to ensure the country's safety is guaranteed.

During the booming prosperity of the '90s, the industrial marketplace was by far the largest contributor to the strong economic growth of the country. The manufacturing sector accounts for an estimated three quarters of all U.S. economic output and 64% of exports. Manufacturers are certainly no stranger to heavy investment in R&D. It is the lifeblood of product development and accelerates time-to-market, which in this day and age is as valuable as anything.[2]

It is really up to the government and other technology providers to help encourage the investment in advanced equipment and tools. A greater awareness of what is available and how small and medium-size manufacturers can access these programs will go a long way in driving technological investments across all the industrial sectors within the United States.

While most SMEs recognize R&D is extremely valuable to many aspects of the production cycle, however, they are the first to admit it is becoming very expensive.

In the pharmaceutical industry alone, the Pharmaceutical Research and Manufacturers Association admits that $32 billion went toward R&D in 2002 and only three out of every 10 companies, on average, will generate enough revenue on a certain product to recoup those research costs. Quantified, that means of every 10 drugs brought to market by a company, seven never turn a true profit.

The reasons for escalating costs associated with the R&D phase vary from heightened government codes and regulations to more complicated and extensive research time. Congress needs to approve a budget that allows small and medium-size companies the ability to access grants that encourage even more manufacturing innovations. A solid commitment in this direction will go a long way to rectifying and even reversing the downward trend.

Despite the continued costs and complexities involved with R&D, it is not a process that can be scaled back. Rather, using more in-depth technology and enhancing collaboration efforts are the best options for trimming both time and expense. It is often very difficult to know if the time and money spent on R&D is providing a solid ROI, but there are definite holes within the R&D cycle that can be filled.

Manufacturers spent all of 2002 and most of 2003 anxiously waiting for an end to the drastic slowdown that accompanied a short recession but was not rebounding quickly enough in the wake of a weak economic recovery.

Many economists believe economic recovery has been on the Bush administration's back burner, with double front burners blazing for an end to Saddam Hussein's reign and the neutralization of what was believed to be a large weapons cache of chemical and biological warfare weapons—weapons of mass destruction.

Just before the war commenced, Commerce Secretary Evans announced a federal initiative to examine factors affecting the global competitiveness of U.S. manufacturing and to recommend ways government can improve business conditions for the struggling industrial sector.

Evans had vowed that the administration's goal has been to identify the challenges facing American manufacturing and outline a strategy for ensuring that the government is doing all it can to create the conditions that will allow manufacturers to maximize their competitiveness and spur economic growth.

Although President Bush might understand the critical need to increase the competitive strength of America's manufacturing sector and create jobs throughout the economy, he needs to work closer with Congress to spur the necessary growth required to jump-start manufacturing. Congress also needs to lend a hand by supporting and passing bills that ease the financial burden on manufacturers. The government needs to create the necessary incentives to encourage manufacturers to reinvest in building America's manufacturing base so it can maintain its world-class status.

6

Action Plan

Throughout the economic hailstorm of the past 24-36 months, manufacturers admittedly have been battered and bruised from all sides. And the outlook for moving forward doesn't look all that much brighter, at least in the short term. Without question, there's continued "rough water" ahead for manufacturers. Manufacturing executives want to be optimistic, but the roller coaster ride they have been on for the past few years is beginning to take its toll.

While this fact might be true, it's time to retune the manufacturing engines and move forward. It is time that Americans learn from their mistakes and help to strengthen the manufacturing sector by looking forward, not backward. It sounds like an oxymoron. But consider what manufacturing has done to strengthen and grow the economy, even during these turbulent times.

Manufacturing has played a pivotal role through innovation, securing our nation, and a solid employment base, just to name a few. In moving forward an action plan needs to be created if the United States intends on reviving and resuscitating American manufacturing. The first step includes new globalization policies.

Second, the U.S. will always have more control over its own actions than over the actions of other countries. Therefore, the key to reversing America's manufacturing decline clearly lies neither in more open-market trade agreements nor in efforts to micromanage economic and social conditions overseas. It lies in establishing fair trade agreements that put all manufacturers on the same level playing field.

Clearly, the first step in addressing the cancer that is destroying the manufacturing industry is by looking at the current trade agreements and improper and perhaps illegal actions of some foreign countries, including some China-based firms.

Despite decades of so-called free trade agreements, too many trading partners still remain closed to U.S. exports. Too many foreign partners routinely establish unfair trade barriers and manipulate currency values. At present, China appears to be the biggest offender in this regard, and is emerging as the primary threat to the industrial sector.

The main reason: most of the world's countries view trade as a zero-sum game, with a piece of the American domestic market as the prize. The handful of economies wealthy enough to consume American-made goods can erect trade barriers faster than U.S. negotiators can even identify them.

We need to act upon the lessons learned. This is a message we should all heed. As business owners and leaders of your corporations you are in the perfect position to formalize a competitive plan to help other manufacturers. Although we have said it before, it's still worth repeating—you need to come together as an industry to help one another. Don't just nod in agreement. It's time to act. It's time to get involved. It's time to help each other and to work together. A collective voice is without question a stronger voice.

There seem to be many reasons, not just one, for the manufacturing crisis. Regardless of the problems, the manufacturing industry is still willing to work together to forge new paths to competitiveness.

EMERGENCY PLAN

In order to begin repairing the damage that has been caused within the industrial sector, the administration needs to set in motion a very strategic plan to resolve the issues impacting the manufacturing marketplace. This might even require the president of the United States to issue a presidential order that declares manufacturing in a state of emergency. While this seems a bit dramatic, the administration needs to focus more attention on restoring domestic manufacturing to world leadership, while ultimately boosting manufacturing employment and wages.

How can the government drive change? It must focus on increasing employment and wages within the manufacturing community. It must issue trade equalization tariffs targeted on countries running large chronic trade surpluses with the U.S.

Declare a moratorium on all current and future trade talks pending development of a new trade strategy that means fair trade for all manufacturers. Given the benefits that China is currently experiencing with the existing trade agreements, there is no real incentive for its government to change.

The Reagan administration is credited with saving a critical Harley-Davidson plant. President Ronald Reagan acted swiftly to protect the American motorcycle industry from damaging Japanese imports by imposing tariffs on Japanese exports. In 1983, two years after 13 Harley-Davidson senior executives signed a letter of intent to purchase the company from AMF, the company was able to convince the International Trade Commission (ITC) for tariff relief against imported Japanese motorcycles 700cc or larger as a response to Japanese motorcycle manufacturers stockpiling inventories of unsold motorcycles in the U.S.[1]

In addition to promoting the right policies at home, the Bush administration needs to exert similar pressure on other nations. A number of countries manipulate the dollar, otherwise known as "competitive undervaluation."

The U.S. government needs to impose stiff tariffs on countries that manipulate currencies for trade advantage. This will quickly level the playing field for all competing manufacturers.

China appears to be one of the biggest culprits of this strategy. China has benefited considerably because there has been a dramatic acceleration of the trade deficit during the past couple of years. As the trade deficit has risen, the value of the yuan, the Chinese currency, has increased simultaneously as market forces would suggest, because its value is tied to the value of the dollar.

While the Bush administration has begun to step up its pressure on China to readjust the value of the yuan, it is not pushing hard enough. So far, the U.S. government has failed to achieve any real success in preventing China and other countries from manipulating its currency.

Washington needs to establish fair trade policies that reduce the U.S. trade deficit, protect U.S. trade laws, and provide for enforceable workers' rights and environmental standards in trade agreements.

What's more, Congress needs to revise tax laws that provide more incentives for corporations to remain in the U.S. and that also discourage manufacturers from closing plants here in order to move overseas. The administration and Congress need to provide tax incentives that help American manufacturers create U.S. jobs and help workers cope with retiree healthcare and pension costs.

While the Chinese leadership and other governments also employ a variety of subsidies and other tactics to help attract manufacturing facilities and boost manufacturing exports, the U.S. government needs to put an end to unfair competitive practices.

We also need legislation that penalizes companies that incorporate overseas to avoid taxes and denies government contracts to these companies.

Some argue that making these adjustments do little if anything at all to correct the problem. However, the right infusion of enthusiasm could spark new life into American manufacturing.

While all of the manufacturing sector's woes are not caused by China and other foreign competitors, it can't be stated enough that much of the intense pressure has come at the hands of the Chinese government and its lack of protecting and enforcing intellectual property rights. The U.S. government needs to do more than wish its problems facing manufacturing will go away.

Both Houses of Congress need to stop political bickering and focus on creating the incentives for American manufacturers. While much of the problems that face American manufacturers are of our own making, if we come together as a nation we can begin the healing that is so desperately needed.

MANUFACTURING TRAINING

To remain globally competitive, education and worker-training strategies must be at the top of the national priority list. Manufacturers clearly support the development of improved vocational/technical training at both secondary and postsecondary levels, as well as programs designed to improve the skills for manufacturing jobs.

For the better part of three years, unemployment has been rising to disappointing levels in the United States. The latest figures in June show the unemployment rate at 5.6%; this is in comparison with 4.3% in March 2001, which was when the last recession began. Yet, many executives are more concerned over the underlying problem that is being somewhat masked by the overall unemployment dilemma: "Mass skills shortage."

Figure 6.1 U.S. **UNEMPLOYMENT**

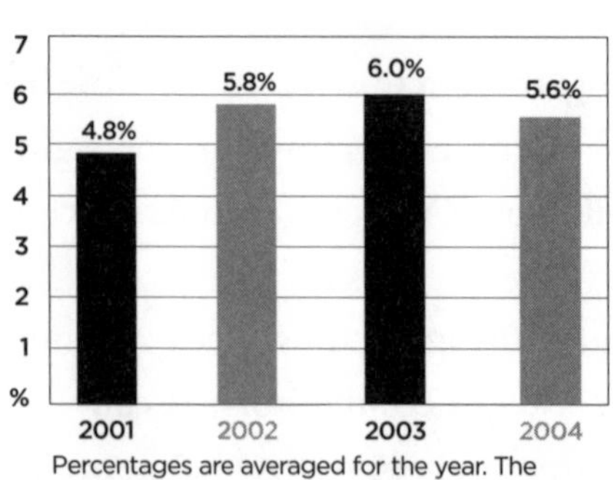

Percentages are averaged for the year. The percentage for 2004 is an average of the first six months.

Source: Start Magazine

There is great concern regarding what many have labeled a "mass skills shortage" in the manufacturing industry. To date, many manufacturers report a moderate to serious shortage of qualified job applicants.

China is reportedly turning out about 300,000 engineers a year compared with 62,000 in the United States and they are technically just as good as U.S. educated students. Therefore, more and more companies are conducting their engineering in China. U.S. manufacturers are finding it more and more difficult to attract young talent, with the aging workforce nearing retirement. This talent void is causing many firms to be concerned and to seek out new ways to deal with a combination of human resources and advanced technology.

Without question, demographic shifts, failures of the educational systems, and an outdated image of manufacturing tied to such stereotypes as the "assembly line" have turned off many young people entering the workforce from jobs in the industry.

Manufacturers agree that a talent shortage could not happen at a worse time, considering manufacturing is already in trouble, struggling through its slowest recovery since the early 1900s. For the most part management has confidence in its employees, particularly those who have moved up through the ranks, say manufacturers. Some manufacturers are huge proponents of promoting from within. This helps familiarize workers with all aspects of the company, which in turn creates an employee base that knows more than just one part of the company. Though this type of philosophy is typical in smaller companies, even large companies with divisions of up to 5,000 workers, are bringing in key talent with the intention of advancing them up through the ranks. Many executives believe maintaining these workers is most critical, and creating an environment that helps them succeed as well as gives them the opportunity to grow is the best option.

Another way to attract new talent is by purchasing advanced equipment that makes it exciting for new employees to learn as opposed to the standard old equipment.

The biggest problem facing small and medium-size companies is combating this mass skills shortage. With the rising costs of healthcare it becomes increasingly more difficult to attract and keep skilled labor on staff. Therefore, for the employee base that has remained loyal for a number of years, the chief concern among executives is the rising cost of providing healthcare insurance.

While a younger labor force has proven to be less loyal than an older more established workforce, the pain occurs when you get someone who has been with you a long time and their contributions are vital to the business, but their healthcare costs skyrocket. It is hard for any small and medium-size company to provide competitive packages until the government steps up to limit costs. This problem will continue to grow to be a much bigger problem than it is today if the government does not take a good long look at rising healthcare and other costs associated with running a small and medium-size firm.

The administration needs to reduce the costs it imposes on manufacturers. The government must take immediate steps to bring down the cost of manufacturing, including regulatory, energy, legal, healthcare, and pension costs.

Most of all, the concern about the skilled-labor shortage is the potential to limit a company's ability to grow. Without the proper people to help improve quality with existing products, it greatly restricts growth into new areas as well. It should be noted that many manufacturers are not as concerned about labor's effect on the bottomline as they are about how labor issues will affect their ability to be accurate, provide top-level service, and produce a consistent quality product to drive sales.

Once the skilled labor issue is addressed, then SMMs or SMEs can look at developing other strategies to find new customers.

In Consultant Collins' opinion, the major challenges are external problems described by analog, not digital, information. The problems are very analog because they are driven by external and analog forces; i.e., customers, competitors, and markets. In order of importance they are as follow:

STAGE 1 – Developing the strategies to find and keep new customers

1. Finding new markets and customers

Customers are dropping SMMs or SMEs sourcing offshore, closing U.S. plants, and moving product lines to foreign countries. Many of them are also forcing SMMs to look for different customers because of relentless pressure for cost reduction. SMMs must react by finding replacements for these customers and new markets for their products and services if they are to survive. However, promoting and finding new customers and markets has never been a strong suit of manufacturing companies. Manufacturers need help.

2. Innovation and new products

To compete, SMMs are going to have to develop more new products and faster than in the past. The success rate of one in seven new products being a commercial success is miserable and they need methodologies to increase their success rate. This has always been a weak point of SMMs because they enjoy winging it on new product development and they do not gather the right external information to ensure there will eventually be sales. They need a lot of help.

3. Changing sales and distribution strategies

Finding new customers and markets is inevitable, but it is forcing many SMMs to rethink their sales channels. Because of rising selling costs and customer demands for more specialized products and services, the traditional industrial sales channels (factory salesforce, distributors, and agents) don't always work. SMMs need to develop a variety of alternative channels to service new customers and market niches. But most are clueless on how to do this and they need help.

4. Adding specialized services

To hang onto a customer account may require more than just developing new products or new sales channels. It will also require developing unique services that bring value to the customer. This can even mean taking over processes that were normally done by the customer. It requires a totally new way of doing business that requires excellent customer contact and external information, which are traditional SMM weaknesses.

5. International marketing

Everybody is coming to the U.S. to take American markets and customers. One strategy U.S. companies have not worked hard enough on (but the time has come) is to go after foreign markets. American manufacturers have traditionally had such a huge opportunity for markets and customers in America that they did not have to look elsewhere. But there may not be enough market niches for everyone, and small and medium-size manufacturers may have to really focus on foreign markets. Industrial marketing is not the strong suit of SMMs so they will need guidance and support.

STAGE 2 - Developing strategic renewal

Once the problems of Stage 1 are solved and manufacturers are assured of having customers and markets in the future and the products and innovative services needed, they can concentrate on the second stage of the problem—a strategic renewal of their companies

6. Changing manufacturing to respond to market needs

This includes lean manufacturing, Six Sigma, quality programs, supplier alliances, supply-chain management, and all of the software solutions coming from the continued investment in information technology. It is true that all SMMs should be employing lean manufacturing, cost reduction, and improved quality techniques just to stay in the game. These are internal problems requiring internal information. They are the solutions mostly understood and embraced by SMMs because SMMs are operation and internally driven companies by nature. This is currently the most popular approach to revitalizing American manufacturing, but it is really the cart before the horse. As an industry we need to invest just as heavily into solving Stage 1 problems as we are doing with Stage 2.

7. Changing the organization

This step automatically follows Steps 1-6.

8. Developing a turnaround plan

This step automatically follows Steps 1-6.

9. Workforce education and training

Another closely related topic with many challenges that suggests there must be a resurgence in education and training to make manufacturing more competitive.

Collins' point is fairly simple. If SMMs cannot successfully solve the problems facing them in Stage 1, then all of the advancements and successes in Stage 2 will be academic and we will not mend American manufacturing.

7

Fighting for Change

U.S. manufacturing is an industry comprised of passionate men and women who are willing to focus their time and attention on sparking new life into the foundation of the U.S. economy. Manufacturers are sounding the rallying cry for small and medium-size companies. Despite the woes and the horrific economic slump for the industry, there is still a sense of pride and commitment to manufacturing. Successful manufacturers faced with a weak economy have been forced to find new ways to become stronger and even more viable during the economic downturn.

And this fact alone is very encouraging. More and more manufacturers are beginning to talk about how they have to change the way they do business in order to survive. Now manufacturers are seeking new avenues for doing business.

Faced with increasingly tough circumstances in the market, many are looking at ways in which to pursue new customers, attack new markets, or simply provide their product and/or service in a manner that manufacturers overseas simply cannot match. In the previous chapters we talk about the need for manufacturers to invest in R&D and technological innovation.

While some American manufacturers are beginning to make progress, we continue to read and listen to news reports about the hardships facing manufacturing. What appears to be even more disheartening is when we learn about yet another small or medium-size manufacturer that has been forced to close his or her doors, or perhaps on the brink of bankruptcy. However, the fact remains, most manufacturers fumble and fail because they haven't adapted to the emergence of a global economy. They are not willing to research new markets or invest in the technology that can make a real difference in their bottomlines.

Other manufacturers admit they aren't necessarily losing jobs to China—they are losing existing and potential customers there. Supplying parts to durable-goods manufacturers and appliance makers—their core business—has been the segment that is primarily moving offshore. To combat this, firms are focusing more on pursuing business in such areas as aerospace and medical equipment, the latter of which is projected to grow quite favorably during the next 10 years. On the other side of the coin, there are some manufacturing companies that are not only surviving, but thriving, because they have taken the appropriate measures to transform their business. They are the ones who are constantly looking for areas of improvement.

Some manufacturers are realizing that they can build a simple product and something that original-equipment manufacturers typically have outsourced to China. These companies realize that most of what they build goes into an end product and they must be able to deliver it very quickly so that it is able to allow their OEMs to be more competitive and turn their product around quicker. In essence they can preserve jobs in the United States rather than outsourcing them to China or to some other region of the world. They have speed and creativity.

Manufacturers admit if they have to compete with India or China on a straight part-to-part basis, they simply can't compete. Thus, U.S. firms must offer better quality and faster turnarounds—something more than just price.

China, while it is creating some serious problems for U.S. manufacturers, is just one component of a much bigger problem. Regardless of the pressures being created, I really think we need to put China in perspective. China is growing very quickly. If it keeps up this pace it will most likely have one of the largest economies before the turn of the century. But according to some estimates, it will take many more years before China can surpass the U.S. It's just not foreseeable that China will surpass U.S. in the immediate future.

And even though China's impact on the world continues to create havoc in many sectors, we should not forget that China is experiencing its own volatility. Rarely do markets skyrocket up or plummet straight down. China's demand for raw materials over the past few years has no doubt contributed to the rise in commodity prices. As a result, China's growth and urbanization is consuming vast amounts of copper and steel.

As noted, the Chinese economy has been growing remarkably fast, yet it is still very small when compared to the U.S. The United States accounts for almost 30% of the world's gross domestic product while China accounts for only 4.4%, about as much as Britain's, but again it needs to be noted that China's economy is growing at a much faster rate than Britain's.[1]

Therefore, if we assume China's economy is climbing at a rate of about 8% per year and the U.S. economy is only growing at an estimated 4% per year then the nominal annual increase in U.S. GDP would be more than three times as large as the increase in China's GDP on a U.S. dollar comparison. Putting it another way, a 1% change in U.S. GDP equals almost the entire annual increase in China's GDP.[2]

Figure 7.1 **GDP BY INDUSTRY**

Value-Added (Real) Percent Changes

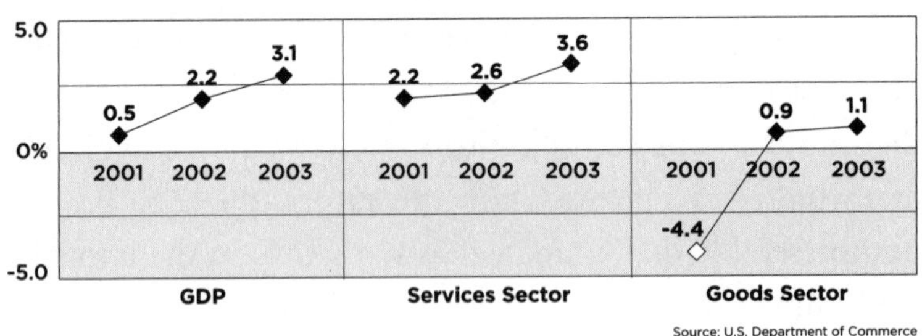

Source: U.S. Department of Commerce

It can't be overlooked that much of China's industrial sector is very dependent on America's demand for its products. Therefore, if the U.S. economy stalls demand for China's products could come to a screeching halt. Thus, China's economy could be severely impacted. If it is true that China already has excess production capacity relative to current demand, China could be headed for some serious problems of its own.

While no one can dismiss China's impact on the world,

it's becoming obvious with these challenges comes even greater opportunities for American producers. Thus, U.S. manufacturers need to be less concerned with China and more focused on boosting American productivity.

Industry insiders believe manufacturing is showing strong signs of beginning to recover, and it will recover because American manufacturers are able to make adjustments and do things differently and quickly when faced with a challenge. Having this ability, whether it is through new strategies or the appropriate technology, will give American companies the upper hand.

Some companies have discovered they need to stay within their niche, such as serving the needs of small to medium-size producers in specific regions of the country. This helps to ensure that they can have almost instant communication with their customers. That is something that many believe foreign competition will never be able to provide simply because they are too far away.

While the aforementioned might not be completely doable for every American manufacturer, the pressure to help these companies find new and innovative ways to compete continues to be a critical factor for determining who will be elected.

Therefore, the only way for manufacturers to really have the greatest impact is by voting for the men and women who are truly committed to helping manufacturing. The only way manufacturers can truly protect themselves is by making sure that elected officials understand the problem and are willing to do something about it. Let's not forget that Americans elected the politicians who serve us in Washington. It's important to cast votes for the individuals who will serve the needs of all manufacturers not just the wishes of a few, large multinationals. It is the responsibility of the manufacturing sector to see that it is electing people who will follow through with the kind of policies that will lead to fair trade.

To date only a handful of politicians really understand the manufacturing crisis. Reps. Ric Keller, (R-Fla.), Don Manzullo (R-Ill.), Tim Ryan (D-Ohio), and Ron Paul (R-Texas) have taken a strong stance to fight to end the decline in manufacturing. In 2003 Manzullo and Ryan even went as far as to create a Manufacturing Caucus in the House in an attempt to help manufacturing rebound.

Most of these political leaders have taken the posture that protectionism is not an economically viable option to revive American manufacturing. The same political leaders insist we need to find new ways to work with foreign countries and not to boycott them. While it's easy to give in to protectionist impulses, manufacturers need to elect officials who are going to fight for change in Washington. We need to elect officials who are going to vote for trade agreements that provide incentives to developing nations to create and enforce higher labor and environmental standards.

The key to rebuilding manufacturing is protecting its capability. Americans need to question the candidates who are up for election this year and make sure they don't have an agenda that differs from that of the manufacturing sector. Americans need to vote for candidates that are going to fight for the betterment of manufacturing.

Manufacturers admit they are very frightened about the future and need some reassurance that their fears are overstated. While many are beginning to recover and things at the moment seem to be brighter, what they are most fearful of is what is out past the horizon.

Macroeconomists love to issue reports telling Americans that this will change into a new economy that will provide us with new well-paying jobs to replace manufacturing jobs, yet no one quite knows from where those jobs are going to come.

Manufacturers need more than just a feeling that the economy will somehow come back and just pull everyone up with it. Still, there are some who believe manufacturing will pull through, just as it always has in the past. But there are others who don't believe just because something has always happened it will happen again.

Although many manufacturers have read articles about manufacturers surviving through these tough times, others comment that they haven't personally observed as many as they would like. The companies they are familiar with have either reduced their workforce or are purchasing parts from overseas.

These same manufacturers insist that something has to be done about the rising costs associated with some of the mandated rules and regulations imposed by the government. And, lastly, they say for American manufacturers to truly compete against foreign competition, they must build better relationships with all members of their supply chain.

During the past 10-15 years, as Americans became enamored with the new, so-called knowledge economy, they somehow lost sight of the fact that manufacturing is the true bedrock of many local economies.

The United States must focus on controlling its own behavior and controlling access to its own market, unilaterally conditioning that access on acceptable practices by its trade partners. In addition, the United States must rely mainly on its own power and leverage it to achieve satisfactory terms of free and fair trade. Washington must also demand that any and all future trade agreements must be reciprocal and strongly enforceable by the U.S. government.

Also, the U.S. government should focus some of its new trade agreements on high-income countries capable of serving as final consumers of U.S. exports.

Finally, Washington must recognize that simply promoting economic growth and higher incomes abroad will not alone cure U.S. manufacturing's ills and rebalance American's trade accounts. Too many countries refuse to trust their economic fates to market forces. Too many refuse to permit higher growth to draw in proportionate volumes of imports. In short, too little commerce around the world is free enough to allow potential future growth to serve as a U.S. trade and manufacturing cure-all.

In order to correct the manufacturing problem, the president and other elected officials must finally acknowledge that the United States faces a manufacturing, R&D, and outsourcing emergency. No longer can the government promote and promise that a recovery is around the corner. Rather, we need to take steps to revitalize U.S. manufacturing competitiveness.

The government also needs to help small and medium-size manufacturers access foreign markets as well as helping them protect their intellectual property rights. However, even before we can begin discussions with developing nations, the government needs to ensure that the dollar is not manipulated.

While the World Trade Organization rules clearly prohibit the use of currency manipulation that would provide an unfair economic advantage, it has ignored the injustices that are underway in China and other countries. With the current political climate, the political pressure is far too overwhelming for the current administration, especially during an election year, to take action to correct the wrongdoing. However, until the current administration takes such action, foreign leaders will not rectify the problems. As we have witnessed in case after case, the Chinese, like other foreign countries, are masters at causing horrific delays within international organizations. And while it appears that China has a long-term strategy to undermine American manufacturing while building up its own industrial base, we need to focus on ourselves and how to revitalize manufacturing.

It cannot be stated enough that there are many contributing factors to the manufacturing crisis. Rather than focusing its attention on China, the U.S. manufacturing industry must learn to successfully manage global businesses. Now manufacturers must spend their time working together to rebuild and strengthen the U.S. industrial sector. The U.S. must become more visible to the world as a leader of quality goods and services.

What some manufacturers have taken away from all the discussions is that they are alone in this struggle. But they are not. Most manufacturers are determined to overcome these hurdles, recognizing the government can be a strong ally if they are educated about the hurdles the industrial sector must overcome. For that very reason most manufacturers realize they must beat a louder drum to rally other businesses to hear the cry of the manufacturing sector. Some outspoken manufacturers are even making recommendations on how to attack the situation more appropriately and swiftly. To keep the economy on track and to become an engine for global growth, the industry must spark manufacturing productivity. To achieve these goals, the entire country must come together to work toward a common goal. Remember working together and raising awareness will go a long way to rectifying the problems facing American manufacturing. It's going to be a tough road to travel. However, Americans have been fraught with obstacles and have overcome; no longer will U.S. manufacturers continue to be an industrial falling star. Now it's time to turn hard times into hot products that generate new revenue streams. At the heart of all of this activity is our ability to successfully unite and fight for change.

8

Manufacturers Speak

DWIGHT CAREY,
President of American Productivity Design and Equipment, Inc.

At last check Dwight Carey has been serving as president of American Productivity Design and Equipment since 1980. The company is a holding company with engineering, manufacturing, distribution, and installation services worldwide, specializing in robotics and factory automation.

One of the most passionate and charismatic individuals today, Carey is very outspoken about helping manufacturers regain their "can-do" attitude in rebuilding American manufacturing.

The 17 companies which he has started and owned have been manufacturing, engineering, and service providers. In addition to sitting on the boards of companies he currently sits on the board of and is a member of "think tanks" which gives him an executive's and an idealistic vantage point which sometimes is quite different from the operational view. A second perspective is that he is a chairman of the Congressional Business Advisory Council and a member of the President's Business Commission. This view is the political, legislative one. Lastly, he is an adjunct professor and teaches a course in strategic management. This gives the theoretical view, but more important, it brings him with the business leaders of tomorrow, observing their brilliance, naivety, shortcomings, and concerns for their future and ours.

He was named Congressional Business Man of the Year in 2003 and has earned the Congressional Order of Merit. He has twice been awarded the Congressional Medal for Business Leadership Technology.

Fascinated with technology, Carey has earned the Technology and Business Award from Specialty Publishing Co., and *Start* magazine for codeveloping a robotic control system.

Why are you no longer a manufacturer after 40 years of entrepreneurship?

First, let me explain that my entrepreneur's clock is still ticking. I was not forced out of manufacturing by one or a combination of the 4,000-plus new regulations our government adds every year to the list of over 90,000 regulations affecting businesses. No, my company was not downsized by competitive globalization. No, I was not put out of business by a class-action lawsuit which stealthily threatens every business, corporate officer, and board member. Nor did my escalating raw materials, labor, insurance, and operating costs wipe out my cashflow. Nope, I am not behind bars because I was wrongfully accused but willfully sentenced for insider trading.

I did buy out my partners and chose to outsource every aspect of my business. The immediate advantage as a manufacturer was that it probably added 10 years to my life, gave me better control of all my direct and indirect costs and put me in a stronger position in guarding quality control. But why take this approach when manufacturing has been in my blood since I first stood at a drill press and assembled small electronic components when I was 11?

Entrepreneurship, when I started my first business at age 22, only enriched my blood and made my heart pound every time I walked into a factory. The reason(s) behind my choice will unfold as I ask and answer the questions below.

What do you see on the economic horizon?

Today is July 21, 2004. All the "talking heads," political pundits, business gurus/publications, academics, and think tank experts are singing "Happy Days Are Here Again." And, all the numbers on the songsheet reflect that tune, which was as wrong in 1932 as it is today. Not that I am saying that statistically they are atonal, even though the numbers for unemployment, new job creations, bankruptcies, private and public debt, "purchasing," etc., do regularly get politically manipulated.

How can we harmonize joyously when the yearly federal budget deficit will be between $550 billion-$700 billion this year (the lower numbers are massaged by unrecorded social security borrowing), our trade deficit is $500 billion per year, our total unfunded debt for the country including private, business, and all governmental is $35 trillion and the worldwide derivatives' (leveraged borrowing) bubble exceeds $100 trillion? In addition, the dollar has dropped 35% of its purchasing power in the last two years, the M3 amount (amount of cash and demands in circulation) has gone up 299% in the last two decades (It rose 270% in the decade of the 1920s before the Great Depression), the inflation rate for all raw materials, energy, fuels, housing and foods is in double digits for the last 12 months (single-digit numbers have been manipulated), and we have two presidential candidates who are debating how much to increase our war efforts and spending, our social spending, and reversing our tax reductions.

All this, and all the world's currencies are backed by a reserve currency, the dollar, which the Federal Reserve has driven down from a 1913 value of $1.00 to three cents today and will further devalue with every new bill printed. These macro issues do not make me want to jump up and sing Hallelujah, no less "Happy Days." No economy ever goes straight up or down. The Great Depression had six major recoveries of as much as 60% between 1929 and 1936. The market did not recover to its 1929 level until 1955. So we will all ride a roller coaster. The severity of its drops will be controlled by the Fed's ignorant and political policies.

What do you see on the international horizon?

By 2050 India will have the world's largest population. It is a country with an English judicial and linguistic history which is cornering the world on computer experts. It is also where IBM is moving 90% of its software development, 90% of its manufacturing, and 90% of its global workforce. It is where more of our X-rays and credit reports are being read, where more Wall Street and service-tech action is happening. It and China are members of the nuclear club.

China, the current villain in the globalization choir, will most likely have the world's largest economy by the turn of this century. It will not be easy. Fifty percent of its bank loans are in default, it needs everything to grow its infrastructure, and it needs an educated population. Already, all our major public companies have or are building factories in China, which is General Motors' second largest market and the home of the only Cadillac plant outside of the U.S. A lot of multinationals are betting that China will succeed. Today China and Japan are holding more than $700 billion in U.S. currency and other financial obligations. The yuan or remnimbi, the Chinese currency, is fixed to the dollar and the Japanese government keeps devaluing the yen so it does not get so strong versus the dollar that it hurts their exports to us. Think of this: they make digital toys and Toyotas and ship them to us, we send them paper; i.e., dollars and treasuries. If they get nervous about the papers' value they will ship it back to us and cause rampant inflation. A very big hammer in trade negotiations.

Recently, we tried to take a tough trade line with China and they only threatened to cancel $6 billion in U.S. soybean exports. We backed down. We imposed a protective tariff on steel and the European Union threatened action against us. We backed down declaring it worked after 200,000 U.S. jobs related to the use of steel were lost to garner 75,000 steelworkers' votes. We tried to bully Canada into holding down the prices of lumber. She retaliated with a tariff. We tried to call in some personal favors to keep oil prices down and no country in or out of OPEC complied. We waged war in Afghanistan to secure a pipeline deal and in Iraq to sit on top of the world's second largest oil reserves.

NAFTA and GATT turned out to be injurious to us and

counter to all the promises made by the politicians and those "in the know." To add insult to injury, under the terms of NAFTA, our government will compensate any companies which would suffer job losses with its enactment. Sounds fair? The top 100 U.S. multinational companies which moved manufacturing to Mexico and eliminated 300,000 jobs will now be compensated for their cutbacks. I certainly did not see negotiating strength, honesty, morality or intelligence from our leaders on the character horizon.

An international question for our politicians: What is the difference between "free trade" and "fair trade"? "Free trade" only occurred from 1815 to approximately 1900. Most all trade barriers came down, commerce, currency, and individuals were allowed to travel freely anywhere in the world. Only Russia and the Ottoman Empire required a visa. Mercantilism and colonialism gradually shriveled. If your company wanted to manufacture and sell something in another country, you negotiated with the buyer directly and set whatever parameters to the deal which you and your customer felt were mutually beneficial...no government bureaucrats, WTO, IMF, GATT, NAFTA, etc., to screw the deal and add staggering costs. Compared with the 20th century, those 85 years were the greatest period of world peace with the least amount of life lost. Why? Because true "free trade" discourages bellicosity. Would the manager of a Wal-Mart store, as part of his company policy, sit on the roof of his store using his customers for target practice? Nor would the manager of Target. Inflation during those 85 years was 0%, with an actual slight deflation in spite of the gigantic gold and silver discoveries during those years.

Why? The world was on a gold standard, China was on a silver standard and we were periodically on a gold or bimetallic footing. Money was hard, indestructible, predictable, dependable and controlled government spending. (During the Civil War Lincoln went off the gold standard and the Confederacy went to a fiat paper currency, which like all paper currencies became worthless as it inflated.) Shortly after the war we returned to hard money. Now, "free trade" as with "fair trade" does not exist today, for different reasons. "Free trade" became politically unpopular with the rise of the omnipotent welfare/warfare state and economic "planning" culminating with Keynesianism in the 1930s. "Fair trade" is oxymoronic and cannot exist in the real world. Who is going to level the "playing field," set the rules of the game, put up the goal posts and where? Will all the teams willingly accept rules which are hurtful and mandated? Will they allow the groundskeeper access to their stadium to destroy their lawn during the cutting? And what if they don't? Ultimately, enforcement as with tariffs, duties, etc., leads to war.

What has history taught you about understanding the political horizon?

The solution to all our problems is very simple but made very difficult to obtain. The problem has been hardened and entrenched for more than 200 years. Please stick with me through this history lesson. Understanding this makes the horizon brilliantly clear and will bring us to my last question and answers. After the American Revolution, there were two economic beliefs in mortal combat. The Hamiltonians wanted the British mercantile system of economic and political favoritism to prevail. It would benefit the north and nascent industries. It would also encourage cronyism, corruption, and a meddling government. Against that, the Jeffersonians wanted a weak central government as established under the Articles of Confederation. A new constitution came out of the first round of that struggle. Agrarianism was not the primary issue with the Jeffersonians. They feared a strong government, which the Americans just defeated in the Revolution. Concentrated power and a central bank would eventually lead to an uncontrollable Leviathan, which would eventually destroy the Republic. Fast-forward 100 years. After the Civil War the government retained much of its power gained in that struggle, began to savor "manifest destiny" spreading west beyond California and acquired a taste for empire.

The Industrial Revolution, along with European investment, foretold that the 20th century would belong to America and its great industries. The only fly in the ointment was that leaders of these growing businesses and fortunes realized that under free-market capitalism there would always be competitors to challenge their supremacy. Business would always be a struggle being won by the next better idea, which would threaten the old order and products. Cronyism and the "best senators and congressmen money could buy" to the rescue. Partnering with government, these titans of "political capitalism" created the progressive movement, social welfare, bought control of the 25 leading newspapers, created the Federal Reserve, instituted the graduated income tax, funded departments of economics, history, and journalism in the Ivy League schools, and even created new universities founded on Hegelian, German principles. The New Deal, Fair Deal, New Frontier, etc., in that soil sprung up as easily as weeds. The old mercantile principles of protection and prosperity for the "haves" morphed into our current one-party Democratic-Republican system of government.

Two quickie questions: What is government and what is it not?

Government is not a cheerleader. Government is not a sugar daddy. Government is not wise, judicious, and all-knowing. Government is not humanitarian. Government is not a producer of capital. Government is not efficient. Government cannot be anything but bureaucratic. Government cannot be run like a business because it lacks the profit motive and accountability, which determines a business's success or failure. Government cannot govern in another government's jurisdiction. Government is the use of coercion to get its way. Government's power to tax ultimately resides in force, which it jealously monopolizes. Government ultimately can wage war for any rationale. Government, when it possesses the power to create paper money, fiat currency, creates economic distortions and uses that paper to perpetuate itself. Government needs "court historians or intellectuals," the media, etc., to sell itself. Have you ever heard a tyrant say, "We are going to invade __________ because I am an immoral rapacious dictator? Every government always develops a logical argument for its actions; to save the children, to save the earth, to save a defenseless friendly nation, 'to make the world safe for democracy', etc." In ancient times the despots may have been more frank by saying, "We are going to rape and plunder your kingdom because you have great looking women and more gold and jewels!" But I cannot verify that. Our government was going to be different because it was founded by people who learned from history why people prosper and how to protect the governed from the government. Nice try, but we have sure slid a long way from the ideal.

Given all the pessimistic scenarios why have you decided the horizon is sunrise for American manufacturing?

Government, nor anything nor anyone, can permanently do anything. No one thing or person can control social outcomes. There are just too many trillions of human interactions to be directed and controlled. Change is inevitable. Those billions of people in the world today acting in their own self-interest with individual desires and intelligence are ultimately uncontrollable. Governments cannot even control their own prisons by keeping drugs, sex, and corruption out, so how will they control the world? Government is the problem, not the answer.

Therefore, what is to be done?

As manufacturers and Americans, demand that your local, state and federal representatives repeal all business regulations which cost American businesses an extra $1 trillion/year. Reverse the 1966 change in the class-action litigation requirements from today's "all people in the class are automatically included in the suit, unless they specifically and in writing request to be excluded from the suit." That sounds unimportant but that is what gave birth to the litigation frenzy, which stalks all of us. "Do nothing and you are automatically a winner in the litigation jackpot." While you are in front of your representative asking for tort reform, ask that we have lawyer compensation similar to the rest of the world...loser pays all legal bills. The "American system" encourages frivolous litigation. You have nothing to lose if you sue everyone and every company within striking distance.

Demand the elimination of all business taxes. The government must live within its means. Eliminate the IRS. America's government and people did fine before 1913 without it.

How will the government provide for "national defense" and the other items granted to it under the Constitution? Read the Constitution for the answer. But the government is too big and needs more money. Ah ha! What about the poor and indigent, etc.? Before World War I there were hundreds of thousands of "friendly societies" and private charities for every race, religion, industry, and institute. These provided more humane and local aid and encouraged people to get back on their feet. Those who honestly could not were cared for by those with an interest in them. Welfare and the War on Poverty institutionalized poverty and the dole.

If your representative does not fight your fight, get rid of him in the next election. Lay the ground rules out to your next representative. Pandering to the special interests of the voter should mean fighting to control spending, putting the dollar on a sound basis (i.e., gold), eliminating debt and refraining from interfering with the natural workings of the market. He or she could become very bored with nothing to do after he/she has won your battles.

Band together with your suppliers and vendors to get lower prices, better products, better terms, better delivery. Form vertically integrated alliances. Association groups should explore self-insurance, larger discounts, and real educational institutions to supply you with the highly skilled and motivated colleagues, who want to help partner with your company's and industry's success.

I decided to outsource everything because I could give to my alliance partners based on their specialized higher skills and talents.

Without the tremendous financial burden of all government regulations and taxes, you can compete with foreign manufacturers. You can go after their markets. Learning to export is not easy for you or them. But you may decide that is part of your niche.

If you have been to the Orient you see it is teeming with consumers, billions and billions of them. There are more 760LI BMWs sold (at $210,000 each) in Shanghai than in the entire United States...more cellphones, bicycles, eventually more of everything. They are also very driven, very disciplined, great savers, very competitive, will make any sacrifices to get a better life and think with a long-term prospective. Just the people you want on your team. Do not write them off as either consumers or team members. That is why all the public and international companies are setting up shop all over Asia and the Muslim world. It is not easy, but they are willing to invest the capital now. In the long run it will be cheaper to do business locally than to import.

The two immediate problems with the developing world are surviving while you rethink and remake your business and dealing with intellectual property theft. That is an international legal issue which you have to wage with the help of good counsel and experts in the field. Costly? Yes. Victory? Maybe. Worth it? Only you can decide on how much pain you are willing to endure.

Again, if you have been to the Orient you have seen many new technologies and exciting products. You have seen the future and I hope have thought about how to prosper there.

Tomorrow's light will start with a sunrise in which every American manufacturer should be willing to bask.

JACK BOLICK

President, Honeywell Process Solutions

Jack Bolick is president of Honeywell Process Solutions. Honeywell is a leading maker of industrial control systems. Prior to assuming his current role in October 2002, Bolick was vice president and general manager of Honeywell Electronic Materials (HEM) in Sunnyvale, Calif. HEM, a business of Honeywell's Specialty Materials group, develops and manufactures a full line of materials used in producing advanced integrated circuits, including solutions based on advanced on-chip interconnect. Candid and open, Bolick talks freely about what he has learned during his distinguished career in manufacturing.

In 1998, Bolick joined Honeywell with more than 20 years of diverse business experience with a focus on semiconductor and manufacturing materials supply, global marketing and manufacturing strategies that support high-growth markets. He was president of Johnson Matthey's Wafer Fabrication business before coming to Honeywell. Between 1980 and 1990, the charismatic and humble Bolick held leadership positions at International Resistive Company (operations director), Analog Devices (industrial engineering manager) and Burlington Industries (quality engineer). Earlier, he was an industrial engineer at United Merchants and Manufacturing.

Earning certification as a Six Sigma Black Belt, Bolick holds a master's degree from North Carolina A&T State University. He also earned a bachelor's degree in industrial engineering from Western Carolina State University in North Carolina.

What do you think is happening to manufacturing?

Manufacturing has been facing unbelievable pressures related to technology. An article I read recently said that in the last 30 years or so in the United States we've been in the final phases of globalization where competition is equaling out from continent to continent. And, according to this article, all of this has brought on a state of "economic marshal law"—economics controls everything.

By the time I was 35-years-old, I had spent over half of my life in manufacturing and seen many changes take place. The first spreadsheet came out in the mid to late 1980s when VisiCalc first came out. Before that, we used paper spreadsheets and relied on mechanical, programmable machines to do the algorithms.

Once we had all the data input on a spreadsheet we would extrapolate information. After VisiCalc, Lotus, and Excel came out, and that's when the leveraged buyouts started. That's when economic marshal law commenced. Suddenly, any accounting or finance person could get data off a mainframe and start crunching data very quickly. They found out a lot of interesting things that way. There were companies just sitting there kind of fat and lazy with tons of real estate that they had to revalue.

That ignited a lot of these leveraged buyouts. Then came the birth of the productivity phase when that same computing power kept getting faster.

There used to be secretaries at every desk, and we made our formal presentations by using the old quarry lettering machines. Then all of a sudden, Harvard Graphics came out, and pretty soon you had a system for one out of every 30 or 40 people instead of having one secretary for every five or six people.

As we moved up the technology curve, more of the "technology" jobs went to Korea and Taiwan. It took Korea about 20 or 30 years to move up the curve, but now, they're as state-of-the-art as we are today and their standard of living is about the same as ours. Now India and China are coming on board.

The difference is those countries produce six times as many high-quality graduates as the U.S. and Germany. The difference right now is that China and India are coming up that curve much faster. Economic marshal law, so to speak, has kept inflation down and created too much capacity—it's caused everybody to hold pricing. We don't have the normal cycles that we used to have.

Now what's happening is everybody has to think of a bigger and better mousetrap. It's not a different mousetrap; it's just bigger and better. Who would have thought that we'd be paying three bucks for a bottle of water?

Was this visionary manufacturing or just marketing?

That was wonderful marketing. Is there really value in that bottle of water versus a beer or a Coke or anything it takes to produce that? No. And yet, we spent years just trying to get from the well to tap water. And you know how good tap water is, even today. What comes out is probably just as good if not better than what is in that bottle of water.

Innovation is the ability to bring manufacturing, marketing, and all those skills to bear. You know how you convince the public to buy? Understanding your customers. What's the value proposition? How does it really have value to help them in this state of economic marshal law? I think that means better education, it means more innovation in R&D, and less tax regulation. I think what's holding us back in this country is our education system.

Students don't perceive there to be many opportunities in manufacturing. So how do we know where to invest our money and get back that edge over the competition? The competition is changing. We've got these big mega markets, totally new economic poles that have to compete globally. I think it's important to keep free trade flowing between them. That's very important. It's important that we not get bogged down in all kinds of regulation.

How is the education system?

I'll give you an example. In the 90s I was up in [Spokane] Washington, and I remember I was really hurting trying to get journeymen. Journeymen machinists have to know a little bit of trigonometry and a little bit of basic calculus. It's a pretty complex trade job and it used to be a very well-respected trade job. You could make 20 bucks an hour. We just couldn't get them. We had to go to the trade schools instead of to the community colleges, but we found that we still had trouble getting people of the right caliber. Local high schools were offering so many different course options that they weren't requiring certain math classes.

Success starts with the family. It's important to pay attention and get students in the right public or private school. The problem is not everybody has that luxury. So I don't know the answer there. How do we get people to go into teaching? How do we get people to know that manufacturing is an honorable trade again? How do we get them back to the basic three Rs? It seems to be a problem in every single state.

Look at our universities today, especially in the engineering and math departments. You'll see how many of those professors are Indian, Korean, and Chinese. The best are coming from overseas to teach here.

Would you then say that more of the problems tend to be our culture vs. the way that students think manufacturing is a greasy job?

I think it is. When we said, "Put a man on the moon," that was a clear directive that the whole country got behind and it drove a lot of technology. Today, I don't think we have that vision. Homeland security is the only thing I can think of right now that's really galvanizing the drive to technology. It's putting money through Congress to go look at manufacturing and find ways we can be more secure. There's not a real positive image of manufacturing. The base of the problem is more the family unit and the education system.

Is the problem globalization or outsourcing?

What got ourselves into this situation were the huge radical technology changes that allowed people to access information and analyze it like never before. That drove consolidation in the manufacturing sector. Companies could no longer live by the ups and downs of the demand cycle. Now you're in this economic marshal law kind of scenario that's accelerated the last 20 or 30 years. There's so much pressure on the leaders of companies and our government to make numbers. You have to ask yourself, Where are the markets? Where am I going? Where can I expand? Globalization is where you can get money. So I think that's what's driving whether or not you're in China. The other side of it is we don't really have an overwhelming vision at the national level saying where we should take this country. The Cold War is over. NASA is not operating as much as it used to. There are a lot of causes out there today. There are things out there that drive our behavior, but I think the biggest things that we have coming ahead are these big economic poles that are starting to develop. What's going to happen in the next phase? Do we draw up curtains around us so it's just America vs. Europe and Asia? You know I don't have an answer for that, but that's the direction it's kind of headed.

Explain why you think it's heading in that direction.

Well, you can see the government regulation. You can see where the pacts, where the alignments, where the lines are being drawn. Now there's tremendous trade, obviously, in resource exchange between different regions but you can see that more restrictions are coming.

How do we get today's students to want to enter the manufacturing arena?

I don't like to say that government is the answer to a lot of things, but I do believe that there should be commonality across all of our wonderful states and territories in terms of education and security. I don't know who else can do it other than the government to step up through a national campaign. Lee Iacocca did it with restoring the Statue of Liberty. We've got to galvanize our population to get a cause. Japan's a good example. They'll come out and say they're going to be No. 1 in the electronics industry. When Ronald Reagan was president, we were at kind of a low point, and he rallied the people to believe we were the best in the world. He was kind of the glue that made it all come together. We need a charismatic leader to step up like that. One thing I will say for sure about Americans—we're winners. You know, we can't stand to lose. And if we focus on something, and we really want to do it, we'll go do it. So it's really just getting that focus.

One of the things that we've said repeatedly in manufacturing, is that if we don't do some things now—and maybe education is a part of it—but if we don't rectify the problems quickly, we could become a Third World country. Would you agree with that?

One of the best things about America is that it's a melting pot. Some of the world's brightest flock to this country to live and work. There has been an overwhelming injection of knowledge in this country to compensate for some of our woes and our poor education. I think if we put barriers up and start shutting our borders, that could be a major problem. We have to have a good environment. We have to make sure we don't just overload on regulations. We just can't stick our head back in the ground like we've done in the past. I think we'll be okay if we don't overregulate, and we start getting some of these things, like education, back on the top of the agenda again. I think manufacturing, all of it will be okay, because the innovation will come, and with innovation will come new jobs.

You don't necessarily agree with or you do agree with the fair trade vs. free trade argument?

I think you have to have free trade, but sometimes in certain cultures, you do have to always have the stick in the background that we have to manage. What we cannot do is just put up barriers and go stick our head in the sand and say we're going to be okay with just the American pole, as these other economic poles start to develop. But they're already developing. I do normally believe in free trade and not only that, you have to manage your borders, you can't just let everybody flood into a country. But at that same time, that's also been part of our strength—the immigration that has happened over the years. You don't want to cut that off completely.

What else should we talk about in the education system, or any other part of the manufacturing crisis we're in?

It is a very complex problem. But I tried to give you some insights, at least, that I believe got us where we are today. Some of us just think this is a natural progression of the economy around the world. The only difference is, it's speeding up. And let's not forget, one of the things that got us in trouble in the '60s for awhile, and we came back from that, was that we bombed Europe and Japan to the ground during World War II. And then we helped them by investing, and we owned a lot of manufacturing companies over there. They helped put the latest and greatest technology into place. And then what happened is, coming into the '70s, we went through a lot of pain because those guys were so much more productive than we were. They had the best factories. What had we done? We had old equipment. Then we went through revitalization.

Everybody said the Japanese miracle was going to roll over us. But we went through a huge revitalization and got ahead of the curve. We moved way ahead of the curve with Microsoft. These are cycles that tend to repeat. We need to look at the past to project the future. We need to continue to invest and not overregulate. Then manufacturing will be okay.

I do believe we need a little more of a plan to getting this noise out of the way and to get education back on the top of the agenda. We tend to wait until there's a crisis to get issues back on the top. But, we also tend to move fairly quickly to fix something.

Anything else we should talk about?

The world's has gotten a lot smaller. We need to work with these economic poles. We should ask, What are things I can do here to make us more productive? We need to constantly shift resources around to match what's happening in the world. There are some companies out there that just shut everything down here and manufacture abroad, but if you notice, that seems to be happening with products that are way down the technology curve.

We're not going to be able to halt this, and if we start regulating like crazy to try, there might be more damage done than good.

CHARLES E. "GUS" WHALEN JR.

President of Warren Featherbone Co.

The CEO of a 121-year-old medium-size infant apparel company, Whalen has seen many changes and survived by adapting to them. Established in 1883, Warren Featherbone has been active over the years in manufacturing, banking, agriculture, and philanthropy. It is one of few companies to operate in three different centuries.

Today Whalen is chairman of the Warren Featherbone Foundation and Alexis PlaySafe Inc., a Warren Featherbone Co., manufacturers of baby apparel. Whalen is passionate and very outspoken that manufacturers need to work together and find new and innovative ways to compete in the United States.

The great-grandson of the founder, Whalen has authored three books while being very active in the company and the manufacturing industry for the past 38 years.

The Featherbone Principle: A Declaration of Interdependence, The Featherbone Spirit, Celebrating Life's Connections, and *The Gift of Renewal*.

He's very active in community and industry affairs-serving as past chairman of the American Apparel Manufacturers Association (AAMA), and former member of the board of directors of Textile/Clothing Technology Corporation (TC2).

Is the manufacturing industry in trouble and why should we care?

The United States has a clear need to rediscover the value of manufacturing and the connection that manufacturing has to the creation of wealth and a high standard of living. As a nation, we have taken manufacturing—as we have our environment—for granted. Most of us know how fragile the environment is, but so too is the U.S. manufacturing sector; and its loss would be devastating. America faces a challenge to maintain a strong manufacturing base. We can meet that challenge.

Many people believe wealth is created in only three ways—through agriculture, extraction (mining, drilling, fishing, etc.), or manufacturing. Beyond those three activities, all other economic activities transfer wealth, but don't create it.

For some time, the United States has been losing its ability to create economic wealth through manufacturing. In the last 20 years, nearly two-thirds of our apparel competitors in childrenswear have gone out of business. In the last 20 years, 18 of 22 major manufacturing sectors as defined by the U.S. Department of Commerce have developed trade deficits. For 2003 the estimated merchandise trade deficit for these 22 manufacturing segments is $410 billion. This is more than $1 billion per day. As this deficit grows, so grows the outflow of wealth.

We are becoming a nation where the rich are getting richer and the middle and lower classes are becoming poorer. We are a more polarized society economically than we've been in many years. In recent years 42% of the wealth in the United States has been held by only 1% of its families. There's an inseparable connection between wealth creation in manufacturing and our standard of living. Manufacturing jobs pay substantially more than service-sector jobs. (2001 average weekly wages in manufacturing were 34% higher than those in the service sector.) With the higher pay comes a higher standard of living, and that standard of living is now declining due to our weakening manufacturing base. In the last 20 years, the median household income has been flat while the cost of living has risen significantly. Today, only 13.1% of total U.S. employment is in manufacturing. In 1979, it was 23.4%. More people are now employed in government jobs at all levels than in all manufacturing segments combined—21.3 million vs. 17.3 million. What this really means is that the U.S. has more people involved in wealth consumption than in wealth creation.

What has caused the decline in manufacturing?

Some of the decline in manufacturing employment is caused by automation and robotics, but more of it has to do with the transfer of manufacturing jobs to other countries (to reduce labor costs) and the growth of lower-paying service jobs in the U.S. I believe one of the reasons many companies are sending so many jobs offshore has to do with a very narrow view of "cost." Component cost is confused with total system cost.

By total system cost, I mean all costs for the manufacturing and distribution of products to the consumer. An apparent lower offshore labor cost may in fact contribute to a higher total system cost due to expenses associated with time and distance. Ironically, we are now seeing some high-tech service-industry jobs like computer programming moving offshore to "low-cost" countries like India. If you think about it, almost everything is "cheaper" somewhere else if component cost is the only cost. It isn't.

What can be done for U.S. manufacturing to make it more competitive?

First, we need to rethink the way manufacturing views itself and how it defines cost. We should look at cost in terms of meeting consumer expectations of quality, price, and responsiveness. Secondly, we need to remind Americans how critical manufacturing is to this country.

As an example, I will use what I know best—Alexis Playsafe, which has initiated changes in operations within its supply chain to meet consumer needs. The apparel supply chain begins with fiber producers and progresses through textile manufacturing, apparel manufacturing, retailing, and ultimately to the consumer.

At first glance, the chain appears to be comprised of separate components, or links. After we analyzed this supply chain along with Dr. Eli Goldratt, business strategist and author of *The Goal*, we became convinced that this chain must be viewed not as separate links but as a total entity.

The term "virtual integration of the supply chain" has been coined by Jan Hammond of the Harvard Business School to describe this alignment, which we call QR2. It means that all members of the supply chain should behave in a way that stresses "until the ultimate consumer buys the product, nobody has sold anything." If virtual integration of the supply chain—QR2 (Quick Response applied across companies)—works for the apparel industry, similar alignments can exist for all other manufacturing segments.

While it is easy to recognize the interrelatedness of members in a supply chain, it is important to consider, also, the interdependence of the various manufacturing segments that might, at first glance, seem totally independent of each other. We have described this interdependence as Quick Response infinity.

Alexis Playsafe and The Boeing Co., can be used as excellent examples. Boeing is the largest U.S. exporter of manufactured goods. As a single company it competes not only with domestic manufacturers, but also with Airbus of Europe. Boeing's power base lies not just within the company but also with the interdependence that it enjoys with thousands of other U.S. companies.

We have common customers and suppliers. We have common problems such as massive consolidation within our industry segments. We have common challenges in supplying consumer demands. We have a common system of government support or nonsupport.

We have common hopes and dreams for our employees. Boeing's prosperity supports ours. Our prosperity supports theirs. The United States as a country will be able to compete internationally only through our ability to leverage this interdependency.

What are you doing to make consumers more aware of manufacturing's contribution?

The Warren Featherbone Foundation was established to increase awareness of the importance of manufacturing in the United States. During 1995, the Foundation cofounded Georgia's first-ever statewide Manufacturing Appreciation Week, which celebrated its 10th anniversary in April of 2004. Georgia is only the 11th state to formally recognize the importance of manufacturing; Minnesota, Michigan, Pennsylvania, Ohio, Illinois, Delaware, South Carolina, Massachusetts, Washington, and Maine are the other 10. This is a sad indication that America takes manufacturing for granted.

We believe that every state in the U.S. should have a statewide celebration for manufacturing and we hope, with this essay, to encourage you to take the initiative in your state.

Here's how the Manufacturing Appreciation Week in Georgia has been developed:

- **A statewide group called "Georgians for Manufacturing" has been organized. It includes manufacturers, educators, chambers of commerce, and state manufacturing groups. A major leadership role is being played by the Georgia Department of Technical and Adult Education in conjunction with the Georgia Department of Education and the University System of Georgia.**

- **A proposed proclamation for Manufacturing Appreciation Week, along with support materials for**

this initiative, was developed by Georgians for Manufacturing and presented to our Governor for his signature.

- **Statewide Manufacturing Appreciation Week has been established annually.**

- **Planning meetings have been held by Georgians for Manufacturing for the sponsorship of Manufacturing Appreciation Week in April. This group's goal is to have a celebration in each of the 497 Georgia communities in which manufacturing activities take place. These grass-roots celebrations will be sponsored by different organizations depending on the community.**

Creating awareness and interest in manufacturing is critical to the long-term economic health of the United States. Manufacturing has provided a substantial amount of wealth to this country in the past and will do so in the future if we recognize its importance and if enough people from all walks of life understand the cost of losing it. It is crucial that our young people learn the importance of manufacturing and the career opportunities it affords. These young people are tomorrow's manufacturing leaders.

Manufacturing Appreciation Week in Georgia and throughout the country is yet another step in greater public awareness that will lead to the rebuilding and mending of our manufacturing base in the United States. Manufacturing is, in fact, an "economic" environmental issue.

We must preserve it and protect it. When we do, we build America.

BRETT KINGSTONE

CEO of Super Vision International

Brett Kingstone is the CEO of Super Vision International, a maker of fiber optic and LED lighting products whose customers include Universal Studios, Walt Disney, and Coca-Cola. The Coca-Cola bottle in Times Square in New York City is lighted with Super Vision lighting products.

Kingstone became an entrepreneur early at the age of 19 while a student at Stanford University. In 1979, with some friends in a dormitory, he designed and manufactured fiber-optic lighting for signs and decorations.

A graduate from Stanford in 1981 with majors in economics and political science, Kingstone founded Super Vision in 1989 and went public in 1994.

Super Vision has been honored as one of the Top 500 Fastest Growing Technology Companies in the United States, Top 50 Fastest Growing Companies in the State of

Florida, Industry of the Year Award in Orange County, and Top 100 Companies for Working Families in Central Florida. Brett is very passionate and an outspoken supporter of rebuilding American manufacturing.

Do you believe China's policy of pegging its currency to the U.S. dollar is hurting U.S. manufacturing? What do you believe is the best way to handle this situation?

Yes, however, I do not believe the United States should dictate China's policy, the U.S. should simply enforce its own policies.

Outsourcing has received much press lately. What are your opinions on manufacturers' outsourcing? Is all outsourcing bad, or is it just inevitable for certain industries?

Outsourcing is bad for the American worker, there is just no way to sugar-coat that fact. Outsourcing has become an "inevitability" for American companies due to the increasingly hostile legal, labor, and tax climate in the U.S. The simple fact is that if our government made it easier to do business here, more manufacturers and more manufacturing jobs would stay here.

What is the best way to spur job growth in manufacturing in the United States?

Enforce our intellectual property laws, place punitive tariffs on rouge nations that violate our laws, encourage retooling and automation in U.S. manufacturing plants, eliminate frivolous lawsuits, and unshackle and empower U.S. companies by reducing regulation and taxation.

Many say the cost of doing business in the U.S. is so high because of healthcare costs, regulatory costs, etc. (costs that many foreign companies don't have to take into consideration) that it makes it very difficult for them to compete. Do you think the U.S. government should consider greater relief for manufacturers?

This is not just something that the government should do for U.S. manufacturers; they should do this for all U.S. companies.

Are stiffer tariffs the answer to what is ailing manufacturing? Why or why not?

Protectionist tariffs will not work, it will only make domestic industries less effective. Punitive tariffs on the other hand are needed against rouge nations that violate our laws as much as jail sentences are needed against criminals.

We must raise the cost of violators of our laws who steal our technology and engage in unfair competition against our companies. This is the one legitimate role of government; however, our government fails terribly in this endeavor. If we create greater costs to infringers, counterfeiters and thieves who steal our jobs, then we will remove the incentive for them to continue to practice behavior with impunity. Right now China and other overseas competitors have far more to gain by stealing our technology than they have to lose. Until this equation is put back in balance, this behavior will continue and so will the loss of U.S. jobs.

Because the U.S. is the epitome of a free enterprise society, there is competition. Other countries don't have this, but U.S. manufacturers have to play by other country's anti-competitive rules when trying to do business there. What is being done to make the playing field more level?

In the U.S., Nothing. Absolutely nothing.

What are some of the positive things you can point to regarding manufacturing?

U.S. creativity, U.S. ingenuity, U.S. technology. These are all byproducts of our free society and our "can-do" American culture.

What do you think needs to be done to rejuvenate manufacturing?

Allow U.S. companies to create, develop, and innovate without interference from government or theft by foreign governments.

How do you think the problems facing American manufacturing can be solved and remedied?

Immediately enforce all our intellectual property laws with severe and meaningful criminal penalties and international punitive tariffs. Eliminate the income tax in favor of a flat tax on consumption. Require all personal injury attorneys to pay the legal fees and penalties for all litigation that they bring that fails to result in a verdict in their favor. Put caps on all personal injury lawsuits to lower insurance and healthcare costs. Provide incentives to companies that invest in automation and retooling to keep jobs and production in the U.S.

There is a lot of discussion about free trade vs. fair trade. How would you respond to manufacturers about what you believe needs to be done?

I believe in free trade. You cannot have free trade unless you have fair trade. We do not have fair trade in the U.S., therefore the only trade that is "free" is the trade allowed to our competitors who sell their products in our markets while not having to follow the same rules and regulations as American businesses. This is crippling to American manufacturing and crippling to our economy.

How can we make vocational opportunities in manufacturing seem appealing to America's youth? What do you think needs to be done?

We need to return to the age-old apprenticeship system where young people who are not inclined to go to college can learn a valuable trade early in life with work-study programs linked to local manufacturers. This will dramatically improve the quality of American labor and, hence, the quality of American products.

What thoughts do you have about manufacturing?

If we continue the path we are on it will become extinct in the United States.

Why do you think the U.S. government is looking the other way when it comes to issues with China?

Our politicians are on their payroll. Chinese political action committees (PACs) are buying too many of our legislators with Chinese money. Also the big wigs at the Fortune 500 companies now have too much invested in their plants overseas and too much to gain from potential future orders in China to allow our government to "rock the boat." As a result we have engaged in a policy of placating and pacifying their violations rather than policing them.

Our politicians have proven that their vote is for sale to the highest bidder, and with it our country's future and sovereignty.

JOSEPH RAGOSTA

President of Oseco, Inc.

Since 1999, Joseph Ragosta has been serving as president of Oseco, a world-class manufacturer of rupture disks and accessories. His products are used to protect pressure vessels such as those used in the chemical or pharmaceutical industries and are used in most countries around the world.

Ragosta has been published in more than 40 technical publications involving product development, material strength, chemical processing, corrosion resistance, and product applications. He holds a doctorate and a master's degree in inorganic chemistry from Cornell University. He is also credited with four patents.

Ragosta is very outspoken, tackling everything from the impact of China on the global economy to the deteriorating effects of the manufacturing sector within the United States.

Do you believe China's policy of pegging its currency to the U.S. dollar is hurting U.S. manufacturing? What do you believe is the best way to handle this situation?

Clearly, China's currency is not properly valued, and they are intentionally doing that to favor Chinese manufacturers. Much of the current trade problem is due to unfair currency valuations. The yuan needs to float against other currencies, as nearly every other major currency does. The market will assign the appropriate value. If China refuses to do so, our government needs to forcefully request it. If they still refuse, we should take appropriate retaliatory action.

We are involved in a war. China realizes it, but we do not. Unless we wake up, they're going to win.

Outsourcing has received much press lately. What are your opinions on manufacturers' outsourcing? Is all outsourcing bad, or is it just inevitable?

There's nothing wrong with outsourcing. Outsourcing is a legitimate way for companies to focus on what they do well and maximize shareholder returns by having someone else do the things that they don't excel at.

The problem is that there's no longer a level playing field. China and India (and a number of other countries around the world) do not have to meet our environmental laws, our humans rights laws, deal with multimillion dollar liability fines for minor claims, or offer benefits to workers and so on. That means that our government is giving our competitors an unfair trade advantage.

Every product imported from outside the U.S. should be levied with a surcharge equal to the cost U.S. manufacturers pay for complying with U.S. environmental and employment laws. Then, and only then, would we have a level playing field.

Of course, we should prepare to pay the same surcharge when selling to countries where the rules are even more strict than ours (parts of Europe). However, the discrepancy between the U.S. and Europe is miniscule compared with the difference between China and the U.S.

What is the best way to spur job growth in manufacturing in the United States?

Eliminate the financial incentives that are currently in place for shifting production overseas. Level the playing field on environmental and employment law issues. America can compete with anyone on a level playing field.

Many say the cost of doing business in the U.S. is so high because of healthcare costs, regulatory costs, etc. Do you think the U.S. government should consider greater relief for manufacturers?

No. Instead, the U.S. should charge imported goods with the cost that a U.S. manufacturer has paid to meet those requirements. This would merely level the playing field.

Are stiffer tariffs the answer to what is ailing manufacturing? Why or why not?

Yes, but only targeted tariffs as described above. The objective should be for American companies to be forced to compete on a level playing field.

How tough can the U.S. be when it comes to the global economy? Because the U.S. is the epitome of a free enterprise society, there is competition. Other countries don't have this, but U.S. manufacturers have to play by other country's anti-competitive rules when trying to do business there. What is being done to make the playing field more level?

As nearly as I can tell, the U.S. is taking almost no action in this area. Our approach seems to be to let foreign governments disrupt the world economy as much as they wish, with no response from us.

To date, none of the major issues have been addressed in any real sense.

What are some of the positive things you can point to regarding manufacturing?

My experience is that productivity in the U.S. is as high as anywhere in the world. Given a level playing field, we can compete, and do it very well.

What do you think needs to be done to rejuvenate manufacturing?

Level the playing field. Put a stop to false barriers erected by foreign countries (such as the bogus pressure-relief equipment rules in place in China).

Force China to let their currency float. The U.S. government and subcontractors should be buying domestic products. Ban goods that are produced in a manner which harms the global climate.

There is a lot of discussion about free trade vs. fair trade? What do you believe needs to be done?

Whoever is making that argument is playing semantic games. Free trade should, by definition, be fair trade.

How can we make vocational opportunities in manufacturing seem appealing to America's youth? What do you think needs to be done?

We can't do anything about this, nor should we. Our government should not be in the position of planning the economy or steering people into a given occupation. If we are able to eliminate the worst of the unfair trade practices, U.S. manufacturing would become more competitive, and we'll find plenty of people to work here.

What thoughts do you have about manufacturing?

Aside from the comments above, American companies are hurt by the need to produce short-term results. In today's business climate, if you can build a plant in China and save $100 million this year, you need to do it, even if you know that you're going to lose access to your technology and lose many times that amount in the long term. We need to find a way to encourage long-term thinking. I'm not sure how you reconcile that with SEC and Wall Street mentalities.

Why do you think the U.S. government is looking the other way when it comes to issues with China?

They're too busy with their confrontational stance in Korea

and the Middle East. This administration is not very amenable to addressing problems openly and peacefully. I would assume that the concern is that if we push China too hard that they will encourage North Korea to misbehave further.

In reality, we hold the cards. If we stopped imports from China for just a few weeks, their economy would be dead. Ours would be hurt, too, but not severely.

HOWARD SANDERS

Vice President of EMF Corporation

Howard Sanders knows a thing or two about manufacturing, having spent most of his adult life in the business, particularly the wiring harness and plastic molding industry. A United States Army veteran who did a tour of duty in Vietnam, Sanders attended Indiana University/Purdue University and Tri State University before settling in the manufacturing sector.

As in the Army where he left with the rank of sergeant in 1968, Sanders worked his way up through the ranks in manufacturing, achieving the status of supervisor, then plant manager, sales, engineering, and finally corporate management where he serves as vice president. He has been with EMF Corp. for 25 years.

Sanders is a firm believer that more can be accomplished to help American manufacturers by electing politicians who foster "free trade" instead of "fair trade," and that foreign markets, particularly China, need our market more than we need theirs.

What can American manufacturers do to ensure the well-being of their companies?

What it really gets down to is we must protect ourselves. And the way we do that is by making sure that our elected officials understand the problem and are willing to do something about it. We elected these people to serve our needs, not the wishes of a few, large multinationals. It is our responsibility to see that we are electing people who will follow through with the kinds of policies that will lead to fair trade. Some people will call that protectionism. So be it. I believe that protecting our manufacturing capability is a necessity and an intrinsic part of what makes the U.S. so great.

We need to question the candidates who are up for election this year and make sure they don't have an agenda of their own but are going to work for us. We are the ones who can shape our government by not voting for those who will continue to foster "free" trade instead of trying to establish "fair" trade. After all, the Chinas of the world need our market a lot more than we need theirs.

Do you believe China's policy of pegging its currency to the U.S. dollar is hurting U.S. manufacturing? What do you believe is the best way to handle this situation?

I have read that China cannot afford to let its currency float since its banking system is about to collapse anyway from bad debts due to government-controlled businesses.

I believe we should make floating currency a condition of most-favored-nation status.

Outsourcing has received much press lately. What are your opinions on manufacturers' outsourcing? Is all outsourcing bad, or is it just inevitable for certain industries?

Outsourcing is going to happen whether it is with China or Mexico or India—wherever there is cheap labor to exploit. As long as we, the American consumer, will accept products made with near slave labor it will continue. I have yet to see where it has really helped the average worker, but it certainly has improved the bottomline of some large corporations.

Do you think the U.S. government should consider greater relief for manufacturers since the costs of doing business keep rising?

These are issues since a prerequisite of being a member of the WTO is that certain human rights are supposed to be met. I believe that our costs are higher because our government forces the human rights issues. I wouldn't want to see them done away with but we need to insist that other trading partners follow the same rules.

Are stiffer tariffs the answer to what is ailing manufacturing? Why or why not?

Yes and no. I believe we should respond in kind. If they want to have high tariffs on imports from the U.S. we should return the favor. After all, most-favored-nation-status must work both ways. If we could remove the barriers to trade it would help.

What are some of the positive things you can point to regarding manufacturing?

American manufacturing is the most productive and innovative in the world. It's where most of the good ideas come from. That must be true or they wouldn't want to steal our designs and methods.

9

Politicians Speak

RON PAUL
Congressman (R-Texas)

Rep. Ron Paul, (R-Texas) isn't your typical politician. He is respected among both his colleagues in Congress and his constituents for his consistent voting record in the House of Representatives. Paul never votes for legislation unless the proposed measure is expressly authorized by the United States Constitution. Former Treasury Secretary William Simon declared Paul as the "one exception to the Gang of 535" on Capitol Hill.

Born and raised in Pittsburgh, Pa., Paul graduated from Gettysburg College and the Duke University School of Medicine, before proudly serving as a flight surgeon in the U.S. Air Force during the 1960s. Paul is a champion for limited constitutional government, low taxes, free markets, and a return to sound monetary policies based on commodity-backed currency. Paul consistently voted to lower or abolish federal taxes, spending, and regulation, and used his House seat to actively promote the return of government to its proper constitutional levels. In 1984, he voluntarily relinquished his House seat to return to his private life as a medical practitioner.

Paul returned to political life in 1997, representing the 14th Congressional District of Texas. He serves on the House of Representative's Financial Services Committee and the International Relations Committee. On the Financial Services Committee, Rep. Paul serves as the vice chairman of the Oversight and Investigations Subcommittee. His priorities include being an advocate for the dramatic reduction in the size of the federal government and a return to constitutional principles.

What do you say to those manufacturers who want the U.S. government to take a protectionist approach when it comes to foreign competition?

First of all, protectionism is not an economically viable option. Governmental activities, including protectionist policies, result in an attempt to have politicians pick winners and losers instead of allowing the natural market process to make these choices. Protectionism results in the government protecting certain politically favored manufacturers at the expense of other manufacturers.

Manufacturers who wish to protect marketshare within the domestic market may see some benefits from protectionism, but those who are attempting to increase marketshare abroad are harmed by protectionist measures. Domestic protectionism results in countervailing tariffs and nontariff barriers. Such a trade war only restricts American manufacturers' access to world markets.

While our domestic market is the richest of any single nation, on the whole, the only way to have sustained economic progress for domestic manufacturers is to improve our global competitiveness.

Simply, we do not represent the bulk of the world market, and as less developed countries make economic progress, exports will be increasingly important to our manufacturing base.

We should also ask from what are we protecting manufacturers? If we truly wish to assist domestic manufacturers, we should protect them from the ever-increasing intrusions of government into the marketplace.

Finally, we should be aware that protectionism is politically impractical. If we look at import policy as a continuum—with protectionism at one extreme and the subsidizing of foreign trade competitors at the other extreme—we see that current policy is actually to subsidize all these competitors.

Far from restricting imports, our current policy has our government taking money from U.S. manufacturers in the form of taxes and giving it to our competitors through various foreign aid and international lending and assistance institutions.

A more realistic approach, and an even far better market mechanism, would be to have our government desist from funding the Export-Import Bank, the International Monetary Fund, and similar entities. It is simply wrong to require American manufacturers subsidize their foreign competitors.

What are your thoughts on NAFTA? Is it working? Why or why not?

I believe NAFTA is not working and that we are setting ourselves up for further failure with the proposed Free Trade Agreement of the Americas, presidential trade authority, and a slew of other similar policy mistakes. Our country has had unprecedented prosperity because of its crucial understanding of the links between freedom, individual choice, and economic well-being. Central to this understanding is our historic policy toward international trade.

NAFTA has very little to do with free trade. At the very base of our understanding of the role of government is the idea that tariffs ought to exist as a revenue-generating activity. For us, tariffs are neither supposed to be used as a method to protect certain favored industries nor as a bargaining chip with foreign governments. Essentially, we believe in the supremacy of market actors over and above political actors.

On the other hand, international trade negotiations and the agreements in which they result imply that government activity is more important to prosperity than are the decisions of market actors. Our trade policy should reflect our own nation's interests and principles, not the fancies of foreign governments and un-American economic theories.

The idea behind giving authority over trade policy to the Congress, that entity designed to be closest to the people, is central to our very system of government.

In turn, the granting of this authority within our system of government reflects our desire to have a trade policy designed for U.S. interests and to protect the freedom of all of our citizens.

Our system, along with the principles it is designed to protect, has been further eroded in recent years. This is the result of two related policy shifts. First, we have shifted our domestic policy from the legislative to the executive branch of our federal government. Similarly, we have created new international bodies, which have the effect of forcing upon us the rewriting of U.S. laws, such as in the WTO Foreign Service Corporation case.

In short, NAFTA is part of a misdirected international system of managed trade that continues the notion of placing primacy in political decisions rather than in the market process. The answer is to unleash creative genius by relying on market forces.

What is the best way to spur job growth in manufacturing in the United States?

The best way to spur job growth in manufacturing is to cut the cost of doing business for domestic manufacturers. This can be done in two ways.

First, we must reduce the crushing burden of federal taxation and regulation placed upon American manufacturers. Real tax reform has not occurred in years, and the only way to accomplish significant reform is to be certain that it includes tax reductions. Increasing government spending results in distortions and inefficiencies, which stifle economic growth, restrain job creation, and hasten job loss. Overhauling our antiquated tax system and greatly reducing the tax burden on manufacturers would be a key step forward.

Along with tax relief and reform, we need a similar serious and sustained effort with regard to regulatory relief and reform. The body of federal regulations continues to grow apace. This continues even as we see outdated regulations that have been around for decades continuing to hamper the marketplace. Regulations breed enforcement agencies, which in turn "vest" in the regulations. Favored industries capture the regulatory agencies as a means of driving up the costs of competitors. All of this causes market-distorting inefficiencies and undercuts the overall competitiveness of domestic manufacturers. It is past time to reduce and reform the regulatory leviathan that has been wreaking havoc upon American industry for far too long.

In addition to adopting tax and regulatory reform, we also need to stop subsidizing our competitors. Foreign aid and foreign trade institutions are heavily underwritten by U.S. taxpayers, not least of all by manufacturers who bear a heavy share of the tax burden through both corporate and personal taxation. In turn, that money is transferred directly through U.S. agencies, and indirectly through international institutions, to entities that compete with U.S. manufacturers in both domestic and foreign markets. This is economic suicide. We must end this policy that is aiding the further dissipation of the American manufacturing sector.

Manufacturers say the cost of doing business in the U.S. is so high because of healthcare costs, regulatory costs, etc. (costs that many foreign companies don't have to take into consideration), that it makes it very difficult for them to compete. Is there any relief the U.S. government can give?

Yes, the United States government can reduce domestic manufacturers' costs of doing business. Again, this is best done through a combination of tax and regulatory relief that can be summed up in the phrase "get the federal government out of the way of manufacturers." This, together with eliminating our financial support for foreign aid and international institutions, would be the way to achieve a significant reduction in U.S. firms' costs of doing business.

This latter idea can be summed up by the phrase "stop subsidizing our competitors."

The important thing to recall is that the reason we ought to undertake these policies is that the interventionist activities that increase the cost of doing business also undercut the competitiveness of domestic manufacturers, thus eroding jobs, economic growth, and standards of living in our country.

Are stiffer tariffs the answer to what is ailing manufacturing? Why or why not?

Tariffs are not the answer for American manufacturing. Again, tariffs are supposed to generate revenue; they are not designed within the American system for the purpose of protecting certain preferred industries. Tariffs lead to trade wars that harm manufacturers who want to increase marketshare abroad in order to achieve sustained economic growth. Foreign markets are critical to many domestic manufacturers.

Tariffs are simply a form of taxation, and so-called "NTBs," or nontariff barriers for the purpose of protecting certain industries are attempts to place primacy over economic activity in political forces rather than in market processes.

We need less government intervention, not more of the same, to reduce distortion. American manufacturers need to look to markets in order to achieve sustained economic growth, and United States policymakers must understand that they can unleash those market forces only by restraining government.

There is a lot of discussion about free trade vs. fair trade? How would you respond to manufacturers about the difference and what you believe needs to be done?

Words like free trade and fair trade have been misappropriated by people abusing the language for political purposes. Fair trade is largely the misnomer that protectionists apply to their discredited idea that the government should increase tariffs and enact nontariff barriers to protect preferred industries. On the other hand, the current policy of foreign trade agreements, presidential trade authority, and supranational international institutions is generally what passes for "free trade." In reality, this policy is more accurately described as "managed trade."

Both "free (managed) trade" and "fair (protectionist) trade" imply government interventionism and hence political meddling in the market process. True free trade would not rely upon any of these notions. In fact, true free trade does not and would not accept these government interventions.

The free market is the result of traders coming together without the interference of government. The sole role for government in such a system is to protect the inherent property rights of the market actors by assuring that nobody engages in force or fraud in completing market transactions. When government ventures beyond these rightful duties it does so as a move away from the free market. In other words, excess government activity diminishes the amount of freedom in the marketplace.

If we apply this same basic idea to international trade we will see why the entire current international trade regime cannot truly pass for "free trade." Free international trade would be a person in country X selling his or her product to a person in country Y at a price and under conditions upon which the two agreed. When international agreements between two governments are put into place or supernational quasi-governmental structures are empowered, politics is given primacy over the market, making the marketplace less free.

Freedom is the absence of governmental interference in the market. It is time we follow a policy of true free trade.

GRANT ALDONAS

U.S. Under Secretary for International Trade

Grant Aldonas is the U.S. under secretary for international trade and has held that position since May 2001. As head of the International Trade Administration (ITA) at the U.S. Department of Commerce, part of Aldonas' job is to advise the secretary of commerce on international trade issues.

Much of what Aldonas is responsible for centers on expanding export opportunities for manufacturers and other businesses and enforcing trade agreements, both of which fall directly in line with many of the issues facing American manufacturers, particularly small and midsize businesses.

Aldonas has been among the more vocal leaders in Washington in an attempt to drive home the plight of American manufacturers, not only to his boss Donald Evans, but all the way to the White House. It's a difficult task for sure.

However, Aldonas has already proven that he can take the heat answering some pretty tough questions from manufacturers. He strongly believes manufacturing's future is strong; however, he insists it will take a lot of hard work not only on the part of Washington, but by manufacturers themselves.

Why should manufacturers re-elect the president?

Until President Bush launched his manufacturing initiative, no administration in history had initiated such a broad review of the challenges facing this sector, and the solutions to help U.S. manufacturers address these challenges. And President Bush initiated this effort, not under pressure from Congress or to score political points. President Bush's concern for the men and women who work in manufacturing and the critical contribution this sector makes to the U.S. economy are the driving forces behind this report.

I am very optimistic about the future of manufacturing in America because this administration is implementing the right policies to ensure U.S. manufacturers succeed. With 759,000 jobs created since August, it is clear that President Bush's pro-growth policies have the economy moving in the right direction.

We are also living in a changing economy today, which means some sectors are producing fewer jobs, but in other areas, jobs are growing. So the president is responding by helping more Americans gain the skills to find good jobs in our new economy. He proposed more than $500 million for his Jobs for the 21st Century initiative to help train U.S. workers for industries that are creating the most new jobs.

A retreat into economic isolationism would jeopardize these jobs, as well as the one in five factory jobs that depend on exports. The answer to outsourcing is to outperform the world, not isolate ourselves from it, and to make sure that American workers have the best training and skills in the world. The president's pro-growth polices are helping American companies and workers to do just that.

President Bush is committed to enhancing government's focus on manufacturing. This includes, for the first time, creating a manufacturing council and creating an assistant secretary for manufacturing and services.

To make the United States an attractive place to invest for manufacturers, President Bush is pursuing an agenda to reduce the costs government imposes on manufacturing. If the president's energy plan is passed, manufacturers will benefit from reductions in natural gas prices, currently the highest in the world. President Bush is also committed to protecting manufacturers from frivolous lawsuits, which can cost manufacturers, even small manufacturers, millions of dollars.

Worker skills and education will be a dominant and decisive factor in America's ability to compete in the global economy. The president is committed to having an education system that teaches the necessary skills to make successful transitions from high school to college and from college to the workforce. If a worker loses his or her job, President Bush is committed to getting them back on their feet by offering personal reemployment accounts, which will help workers rapidly reenter the work force with the training they need for the jobs of the 21st century.

The president's policies have brought the country back from recession to recovery. I believe America will want to go forward with pro-growth policies that are creating jobs and making our economy strong, not change the policies that are strengthening our economic recovery.

Cite some examples of why American manufacturers should be optimistic about their future?

First and foremost, President George Bush understands the importance of manufacturing to our economy, our workforce, and to our future—and he is working to implement policies that will help our manufacturers succeed. President Bush understands that fair trade and expanded exports are vital to our nation's economic strength. In 2004, American companies are selling computer chips to Japan, producing BMWs for export to Germany, exporting California wine to France, and even selling Mexican food to Mexico. When 95% of the potential customers for American products live outside the U.S., America must reject policies that would result in economic isolationism, endanger our economic recovery, cost U.S. workers jobs, lead to higher prices for American consumers, and put U.S. workers and companies at a competitive disadvantage.

The manufacturing sector absorbed significant job losses after the recession, the uncertainties of war, corporate scandals, and the attacks of Sept. 11, 2001. Fortunately, President Bush's pro-growth policies have helped lift the U.S. economy out of recession. In March 2004, hiring in manufacturing increased for the first time in 43 months. The United States remains the world's manufacturing leader, in large part because Americans are the world's most efficient workers. If our manufacturing sector stood by itself, it would be the fifth-largest economy in the world—larger than the entire economy of China.

There are additional positive signs for the future of U.S. manufacturing. The Institute for Supply Management's manufacturing index for the month of March rose to 62.5

on news of increased factory production, the 11th straight month the index has exceeded 50. When the index is more than 50, it is a sign of economic and job expansion.

After-tax profits of U.S. manufacturers were $79.6 billion in the fourth quarter of 2003, up $9.5 billion from the previous quarter and up $45.6 billion from one year ago. Factory activity is now at its highest level in 20 years and new orders are at their highest level since 1950. Most important, the economy has been growing for seven months in 2004—during which it added more than 750,000 jobs. It is critical to continue growing the economy so American manufacturers can benefit from this economic momentum.

What can small to midsize manufacturers do to compete on a more global scale?

The numbers are telling—small businesses are already succeeding in exporting to China. The number of known small and medium-size enterprises (SMEs) that exported to China in 2001 totaled 12,880, up from 3,143 SMEs in 1992. Very small companies—i.e., those with fewer than 20 employees—made up 45% (nearly half) of all U.S. firms exporting to China in 2001. This is up from a 38% share in 1992. SMEs are known to have exported goods to China worth $5.25 billion in 2001. China was the seventh largest market for U.S. merchandise exports from SMEs.

Increasingly, it will be important for more small businesses to find a way to integrate into global supply chains. This may not necessarily mean exporting themselves, but producing for a large company, such as Ford, which is expanding its global exports. This is where the Manufacturing Extension Partnership (MEP) program can help. The

Department of Commerce's Foreign and Commercial Service (FCS) offers a worldwide network of 107 domestic offices and 85 overseas posts to help small and medium businesses compete and win in the global marketplace. FCS staff provides an array of practical services to teach American businesses how to break into foreign markets. Whether a company is new to exporting or is looking to expand into additional markets, the FCS provides effective assistance and timely solutions.

Under the direction of Secretary of Commerce Don Evans, the department has increased coordination of our trade policy and promotional efforts and we are implementing a National Export Strategy that will streamline bureaucracies and focus our agencies on continuous program coordination and improvement. U.S. companies now benefit from better governmentwide support in negotiating more open markets, including aggressive export promotion strategies for each of the free trade agreements (FTAs). Small and medium-size manufacturers interested in exporting should visit www.export.gov for exporting information.

If a rural manufacturer would like to expand into parts production, what government programs (grants, loans, financing, etc.) are available?

For any small business, whether getting started or expanding in a new direction, an important first step is to seek assistance from the Small Business Administration (SBA). At SBA, the federal government sponsors its own public venture-capital organization through the Small Business Investment Companies (SBIC) program. This program can be a key resource for rural manufacturers. During the last fiscal year, more than 25% of SBIC financing funds, or $728 million, was reported in the manufacturing industry. Using a combination of private funds and funds borrowed from the federal government, SBICs provide equity capital, long-term loans and management assistance to eligible small businesses. An SBIC is a source of financing for a fast-growing, existing business that needs a substantial amount of financing to keep up with its rapid expansion. Additional information can be found at www.sba.gov/.

The Manufacturing Extension Partnership program at the U.S. Department of Commerce also can be a valuable resource for rural manufacturers looking to be export-ready and compete in the global marketplace. The department's National Institute of Standards and Technology (NIST) oversees this nationwide network of hundreds of not-for-profit centers working to provide small and medium-sized manufacturers with the help they need to succeed, whether it is restructuring, integrating into the global supply chain, or meeting technical standards. MEP centers serve all 50 states and are funded by federal, state, local, and private resources. The centers can help small firms overcome barriers to relocating in rural areas and assist them with obtaining private-sector resources. www.mep.nist.gov.

What is the status of the government's efforts in working with other countries so that the Free Trade Agreement is more fair trade?

Small and medium-size businesses will benefit directly from new trade agreements that slash foreign tariffs and remove the barriers that disadvantage American workers and exporters. At the beginning of this year, U.S. companies were able to start taking advantage of FTAs with Chile and Singapore, and we have recently completed negotiations with Morocco, Australia, and a group of countries in Central America. The numbers tell the story of how fair and open trade will create jobs here at home, while helping the small business sector of America:

- **More than 6,000 small and medium-size businesses export to Chile.**
- **More than 4,000 small and medium-size businesses export to Costa Rica.**
- **Approximately 3,000 businesses export to Honduras.**

The Bush administration is committed to ensuring a level playing field for U.S. companies competing in the global economy. To this end, we continue our efforts to open new markets and lower trade barriers through our FTA negotiations with four Andean countries, Bahrain, Panama, Thailand, and five member countries of the South African Customs Union. Further, we are working with our neighbors in the Western Hemisphere to create a Free Trade Area of the Americas (FTAA) that will form the world's largest free market.

In the context of the World Trade Organization (WTO), this administration is dedicated to moving forward with talks in the Doha Development Agenda. We still believe that these negotiations offer the best chance to bring economic rewards to U.S. workers and exporters, and to countries and peoples around the globe.

Moreover, the Bush administration has done more than any previous administration to negotiate labor and environmental standards into trade agreements that encourage other countries to institute labor and environmental practices consistent with those in the U.S. market. This helps level the playing field for U.S. manufacturers in the global marketplace.

Please list what actions you plan to take to improve protection of intellectual property.

Intellectual property industries play a significant role in the American economy, contributing approximately 5% to the GDP, employing millions of people, and representing the fastest-growing sector of the economy. While the value of intellectual property increases, so does the ease in which copyrighted material can be infringed at low cost.

Well-organized criminal enterprises have recently increased the scale, scope and sophistication of international piracy and counterfeiting. These individuals can now disseminate millions of copies of stolen software, music, video, and other manufactured products and programs around the world with a simple computer click. With the inconsistent enforcement of existing laws worldwide, it is imperative that intellectual property rights (IPR) be reaffirmed and vigorously protected.

On March 31, 2004, the U.S. Department of Justice created an intellectual property task force that will explore ways to improve protection of patents, trademarks, copyrights, and other forms of IP. The task force will determine how best to meet the evolving challenges that law enforcement faces in the intellectual property arena.

In addition, the attorney general expanded the Computer Hacking and Intellectual Property (CHIP) units. The Computer Crime and Intellectual Property Section of the department's criminal division was also provided additional resources to fight piracy.

China's government understands the importance of IPR protections, but the problems still remain. The Business Software Alliance estimates that software piracy rates in China exceed 90%. It is reasonable to assume that the vast majority of the Chinese government is operating with pirated software, which would be worth $2 billion per year if legally purchased. This is certainly uneven and unacceptable trade.

At the Department of Commerce, three new offices have been created that target unfair trade practices and enforce our U.S. trade agreements. Our Unfair Trade Practices Task Force is already tracking the 30 largest categories of Chinese imports.

We are strengthening IPR enforcement with a new Investigations Office to drill down on the countries tolerating the theft of American products.

For example, Chinese manufacturers are copying Toro's designs and producing identical knock-offs that differ only by adding a "K" before the word Toro.

We are aggressively confronting any country that tilts the playing field against American manufacturers. Whether that country is China or India or any other trading partner, we are cracking down on unfair trade practices not only after they have happened, but also as they happen. Free trade requires fair rules and U.S. Department of Commerce Secretary Donald Evans has taken that message back to China this past summer.

Manufacturers have been calling for a blueprint for action to address concerns of manufacturers. Is the recent Department of Commerce's report on manufacturing this "blueprint?"

The report, "Manufacturing in America," represents President Bush's blueprint for creating an environment where entrepreneurs want to invest in manufacturing. While we are at the beginning of the process, manufacturers can be assured that we will not rest until we have created the economic environment that U.S. manufacturers need to succeed. In developing the report and recommendations, we held roundtable discussions with manufacturers in the aerospace, auto, semiconductor, and pharmaceutical sectors, among others, in more than 20 cities across the United States—from North Carolina to Columbus, Ohio, to Detroit, to Los Angeles.

What we heard from manufacturers in terms of the challenges they face was significant. While international competition has garnered most of the attention in the media, by far the greater weight of the manufacturers' comments focused on domestic issues—what I call "keeping our side of the street clean." What I mean by that is simply paying attention to the needs of our manufacturers as we develop legislation or implement regulations. It is the steady accumulation of multiple burdens, rather than a single cause, which has had the most severe impact on the competitive environment in which our manufacturers operate.

The list of issues our manufacturers identified should not surprise anyone who has taken a serious interest in manufacturing. While they have tightened their belts and raised their productivity in an effort to remain competitive and, in fact, to succeed in the day-to-day competition in the marketplace, they have seen that advantage and the hard-won productivity gains eroded by everything from higher energy costs to higher medical and pension costs to higher insurance costs due to a runaway tort system.

Government has to create conditions that will allow small and medium-size manufacturers to succeed. Fostering an economic environment that attracts investment in manufacturing is key to maintaining the United States' position as the world's leader in manufacturing. And, the key to attracting that investment is ensuring that we offer a competitive environment from which manufacturers can produce, not just for the U.S. market, but also globally.

This report includes a series of more than 50 recommendations along these lines that are aimed at unleashing the full potential of American manufacturers. It is a critical first step toward strengthening American manufacturing and creating new jobs.

What is the next step for this report? Who is in charge to make sure the things proposed get reviewed or enacted?

Secretary Evans is the principal advocate for manufacturing in the administration—a role he takes very seriously. It was under his direction that the review of our manufacturing industry was initiated and he is driving the report's implementation. In the Department of Commerce and across the Bush administration, we are already implementing the recommendations in the report that fall within our statutory authority.

These efforts include establishing an unfair trade practices task force to track, detect, and confront unfair competition before it injures an industry here at home. Its goal is to focus on those trading practices that are likely to have the biggest impact on our manufacturers and ensure that they are eliminated, rather than leaving small and medium-size manufacturers in the United States with costly trade litigation as the only possible means of addressing the unfair trade practices they face in the marketplace.

In April, President Bush nominated carpet manufacturer Al Frink as the new assistant secretary of commerce for manufacturing and services to serve as the point person in the administration and within the U.S. government for manufacturers and as an effective advocate for the manufacturing sector's competitiveness. And Mississippian Rhonda Keenum has just been confirmed by the Senate as the new assistant secretary for trade promotion to boost our exports, particularly to those markets that our negotiators have recently opened to our trade like China.

However, much of the agenda outlined in the report is legislative, and a good deal of work remains to be done by Congress. Last year, the House of Representatives passed many of the key initiatives in the report—such as tort reform and an energy plan—but the bills stalled in the Senate. Currently, the White House and Dept. of Commerce are urging Congress to pass legislation outlined in the report, such as comprehensive energy legislation to create a more affordable energy supply, tort reform to protect manufacturers from frivolous lawsuits, as well as healthcare reform, and association health plans.

We are also working with Capitol Hill to reduce the costs of regulation and legislation on manufacturers. The Bush administration has slowed the increase in regulatory costs produced by new regulations reviewed by the Office of Management and Budget by 70% compared to the previous administration. Nonetheless, overall the cost of regulatory compliance has risen significantly over time.

While I am currently overseeing the implementation of the manufacturing initiative, these responsibilities will be passed to the new assistant secretary as soon as his nomination is confirmed by the Senate.

MURRAY SABRIN
Professor and Former Political Candidate

Murray Sabrin is professor of finance in the School of Administration and Business, Ramapo College of New Jersey, and executive director of the Center for Business and Public Policy.

Sabrin was the 1997 Libertarian Party gubernatorial candidate in New Jersey, and the first third-party candidate to receive matching funds and participate in the official debates. He received more than 114,000 votes, approximately 5% of the turnout. In January 1999 he rejoined the Republican Party and formed an exploratory committee to seek the U.S. Senate seat being vacated by Frank Lautenberg. In the June 2000 primary, he received nearly 13% of the vote in a field of four candidates

He has a Ph.D. in economic geography from Rutgers University, an MA in social studies education from Lehman

College and a BA in history, geography and social studies education. Sabrin has worked in commercial real estate sales and marketing, personal portfolio management, and economic research. He began his career as a New York City public school teacher in 1968.

What do you say to those manufacturers who want the U.S. government to take a protectionist approach when it comes to foreign competition?

Protectionist measures can provide manufacturers with short-term benefits. It is tempting for manufacturers—and in some cases irresistible—to call for more government intervention to protect an industry or sector of the economy from foreign competition. Initially, protectionist measures such as higher tariffs, quotas on imported goods, subsidies to U.S. manufacturers and other policies could boost sales of U.S. goods producers, increase employment, and boost their profits. However, these benefits are not sustainable and, in fact, are harmful to both the industry and their employees. After the protectionist measures are ended—and they always are terminated—the industry may have excessive capacity to meet the needs of the marketplace because they expanded during the protectionist period, and the readjustment will be harmful to both business owners and their employees as they are forced to downsize. Furthermore, foreign countries will retaliate against U.S. companies if their businesses are hurt by the federal government's protectionist policies. Thus, while manufacturers may benefit in the short run, other U.S. companies will be hurt in both the short-run and long-term, thereby canceling out any benefits that may have occurred because of protectionism. In addition, American consumers will be negatively affected as well, because they will have to pay higher prices for manufactured goods. In short, there are no long-term "winners" when a nation goes down the protectionist road, except for the federal government, which now has more power over the economy.

Business owners are in it for the long run. Politicians only care about the short-term; that is, the next election. Business owners and executives should not propose and agitate for short-term—and failed—government "fixes," thereby diverting their focus and energy and creativity away from their primary activity, serving customers better. Manufacturers should advocate good economic polices. That is, they should be in the forefront of calling for litigation reform, regulatory reform, less taxes, lower government spending, and sound money. All these measures are necessary for sustainable prosperity at home, and will make U.S. companies more competitive in the global marketplace.

Manufacturers say China's policy of pegging its currency to the U.S. dollar is hurting them. What do you believe is the best way to handle this situation?

According to *The Economist's* "Big Mac" index, the Chinese currency, the yuan, is about 50% undervalued against the U.S. dollar. The index measures the purchasing power of a nation's currency against a Big Mac, one of the world's most homogeneous products. If a Big Mac is priced $3.00 in the United States, for example, and its price in Europe is 3 Euros, then the dollar should be worth one Euro, making the exchange rate $1.00 = 1 Euro. However, if the Euro is trading in foreign exchange markets at $1.25, then we can conclude that the Euro is overvalued. Conversely, if the

Euro is trading for 90 cents, then it is undervalued.

Because the Chinese government "pegs" the yuan to the U.S. dollar at the rate of 8.2 to the dollar, even though the Big Mac index suggest the yuan should be about 4 yuan to the dollar, Chinese exports are "underpriced" in global markets. Conversely, the Chinese government's peg of the yuan also means foreign goods are more expensive to the Chinese because the yuan is undervalued, thus hurting Chinese consumers.

Can anything be done to "level" the playing field? The answer is no. We are witnessing the inevitable consequences of government meddling in a nation's monetary affairs. As long as governments have control of the monetary printing presses, they will try to do their best to benefit domestic interests at the expense of the general public. A weak currency policy will be a boon to exporters who find their goods more competitive in the global economy. A strong currency, on the other hand, means greater purchasing power for a nation's citizens, thereby dampening inflationary pressure. However, it also means that exporters, particularly manufacturers, could be at a competitive disadvantage in the global marketplace.

As in many areas, government intervention distorts the economy, and in the monetary arena, distorts global trade patterns, undermining a harmonious world trade system.

To meet the challenge of Chinese competition, U.S. manufacturers can restructure their production in order to meet head-on their Asian competitors, and/or they can form joint ventures with overseas companies in China or other low-cost countries. Or, they can move their production facilities to China or other nations in Asia, South America, or Eastern Europe. The choices may not be what many manufacturers want to hear, but the reality of the global economy is that the rules are constantly changing, but as long as governments control money, there will be distortions in international exchange rates.

Only international money divorced from government manipulation will level the global competitive marketplace and produce sustainable prosperity. Until we have a real international gold standard, a monetary system that cannot be manipulated by any government to give any sector of the economy undue advantage, governments will continue to intervene in foreign exchange markets and set polices that will create both winners and losers. The current fiat paper money system is the reason we have international economic tensions.

Outsourcing has received much press lately. What are your opinions on manufacturers' outsourcing? Is all outsourcing bad, or is it just inevitable for certain industries?

First, let's state the obvious. Every family is engaged in outsourcing. That is, every family makes purchases from both domestic and foreign producers and suppliers of goods and services. Every family is dependent upon others to provide them with high-quality goods; families purchase services from "outside" sources they cannot provide for themselves. Most families need doctors, lawyers, accountants, plumbers, electricians, and others to lead more productive lives. In short, the division of labor is a great boon for the human race, as every member of society can specialize in some activity leading to a more productive life.

For most manufacturers, outsourcing is a logical means

to an end...producing low-cost quality products. If U.S. manufacturers, however, can obtain all the inputs for their enterprises domestically, they will of course "Buy American." Thus, basic business decisions have to be left to the discretion of entrepreneurs, who know what is the best "recipe" for their businesses.

Moreover, foreign companies also "outsource" to the United States. Japanese automobile companies, German, Swiss, French, British, Dutch, Italian, Korean, and companies from around the world have established production facilities in America. What some policy analysts and public officials complain about is U.S. companies moving some of their production overseas. They do not seem to mind foreign companies relocating in the Untied States.

Politicians are the last ones who should be deciding whether companies or industries can or should outsource. This is not only a fundamental business issue, it is also a basic philosophical issue. Who should decide how to allocate assets and property in a free society? The answer is obvious: property owners, not the political class.

The cost of doing business in the U.S. appears to be creating more hardships for manufacturers. With the rise of healthcare costs, regulatory costs, etc. (costs that many foreign companies don't have to take into consideration), manufacturers insist it's almost too difficult for them to compete. Is there any relief the U.S. government can give?

The United States economy needs a heavy dose of deregulation, an overhaul of the tax code, and less government spending. The total costs of regulation reached $869 billion in 2002. While some of these regulations could be justified in protecting the safety and security of the American people, the burden of these costs is in effect a tax on every American; in short, a destroyer of jobs. Government mandates need to be reviewed, especially healthcare mandates on employers that drive up the costs of employment. These mandates are, in the final analysis, taxes on hiring workers.

Businesses need to have flexibility in providing benefits to their employees. Health savings accounts are a good first step. Giving employees more control over their healthcare benefits will lower costs for employers.

There should also be an overhaul of the sexual harassment and discriminatory statutes. They put a huge burden on businesses and punish innocent shareholders for the acts of managers and corporate executives. If they are guilty of violating the law, they should be punished, not shareholders. "Deep pockets" do not justify punishing shareholders.

The federal tax code should be overhauled. Reducing the tax on dividends and capital was an important reform, because it virtually eliminates the "double taxation" of business income. A more efficient tax reform would be the elimination of the corporate income tax and terminating the Alternative Minimum Tax (AMT).

All taxes on business income should be paid by shareholders. This would be a great boon to businesses. It would reduce the need for tax accounting services, freeing up resources that would allow firms to concentrate on their core business activities. In other words, all business entities should be organized like S corporations for tax purposes.

In a similar vein, reducing federal spending would also free up resources that could be used by America's business sector to be more competitive in the global economy. Less spending, less government debt, lower tax rates on businesses, if the corporate income tax remains, would do more for the business community than any of the short-term gimmicks—subsidies, grants, loan guarantees, and other interventions—the federal government could implement to "help" America businesses.

How can we make vocational opportunities in manufacturing seem appealing to America's youth? What do you think needs to be done?

For many decades we have been preaching the benefits of a college education as the bee-all and end-all for the best career opportunities. Although we have become a "knowledge-based" economy, a shift that has been evolving for the past four decades, America's manufacturing sector will need not only replacement workers in key industries in the years ahead, they will also need more workers as the economy grows over the next decades.

To attract a sufficient supply of highly motivated workers to their firms, business owners and corporate recruiters need to be proactive in their communities. They need to attend high schools, community colleges and possibly even college job fairs, to highlight the opportunities in their businesses. Manufacturing firms need to "market" their companies. Virtually all young people believe that success means being a white collar worker. I would recommend that manufacturing companies organize tours of their plants. Tours should start with elementary students and junior high school students. In addition, companies could offer internships in the summer for motivated high school and college students.

Manufacturers can also work with local and county work development agencies to attract a pool of highly motivated workers looking for work or seeking a more challenging opportunity.

In short, there are ample opportunities for manufacturing firms to address any pending labor shortages by targeting potential new workers in their communities. The president or CEO of the company or a vice president in charge of production would be the ideal spokesperson to get young people excited about the prospects of working in the manufacturing sector.

DONALD MANZULLO
Congressman (R-Ill.)

In 2004, Rep. Donald Manzullo (R-Ill.) is continuing his mission to restore manufacturing in America and put people back to work in northern Illinois and throughout the nation. He has held more than 60 hearings on the state of manufacturing and introduced related pieces of legislation since his colleagues appointed him chairman of the U.S. House Committee on Small Business in 2001. He also founded the 80-member House Manufacturing Caucus, which he chairs, and he serves on the House Financial Services Committee. Manzullo represents the 16th Congressional District of Illinois, which includes the counties of Winnebago, Boone, Stephenson, JoDaviess, Ogle, Carroll, the majority of McHenry County and parts of DeKalb and Whiteside counties.

Manzullo, elected in 1992, has earned the reputation as

Congress' champion of manufacturing and a fierce advocate for job creation. His "Agenda to Restore Manufacturing in America" outlines 17 priorities to preserve U.S. manufacturing and put Americans back to work. The plan's highlights include providing tax relief to companies which keep jobs in America, forcing China and the other East Asian countries to stop manipulating their currencies to give themselves an unfair cost advantage over American companies, requiring the federal government to comply with Buy American laws, reducing the surging cost of healthcare, expanding U.S. Small Business Administration programs to small manufacturers, reforming U.S. export control policy, and many others.

One of the priorities in the plan, rescinding the steel tariffs, was accomplished in December 2003 when President Bush removed the duties on imported foreign steel, which were pummeling U.S. manufacturers who were forced to pay higher prices for steel than their foreign competitors. Another priority, encouraging job creation in America, is nearing accomplishment as Congress rewrites corporate tax law to avoid sanctions from the World Trade Organization. Manzullo leads the fight in Congress to include as many changes as possible to benefit domestic manufacturing and encourage companies to keep work in America.

Manufacturers say China's policy of pegging its currency to the U.S. dollar is hurting them. What do you believe is the best way to handle this situation?

Treasury Secretary John Snow, Commerce Secretary Donald Evans, and President George Bush were on the right track when they became the first administration to challenge

China's currency manipulation practices during their visits to China, which resulted in more orders for American companies. While progress is being made, China is not moving fast enough to end its unfair practices. The U.S. government should ratchet up the pressure even more to convince China to freely float its currency.

In addition, the federal government should strengthen its trade-enforcement laws to provide more relief for America's manufacturers. I am a co-sponsor of H.R. 3716, a bill that would allow the U.S. to pursue countervailing duty cases against "nonmarket" economies, including China. The legislation, authored by Rep. Phil English (R-Pa.), would allow the U.S. government to impose tariffs on imports equal to the subsidies a "nonmarket" government provides to its exporters. I will continue to work to force China to stop manipulating its currency and level the playing field for U.S. manufacturers.

What is the best way to spur job growth in manufacturing in the United States?

First, we have to make U.S. manufacturers more competitive. Taxes are too high, healthcare costs are skyrocketing, and regulatory requirements are costly and burdensome. In Congress, we have been working on a competitiveness agenda designed to lower the cost of doing business in the United States so our companies can better compete in the global marketplace.

We must also continue to hold foreign countries accountable for their unfair trade practices. China, for example, gets up to a 40% cost advantage over the United States simply because it pegs its currency to the U.S. dollar.

Finally, we must reward U.S. manufacturers for doing business at home. The pressure for low-cost production has lured many U.S. companies overseas, sending many Americans to the unemployment lines. The House recently passed a tax relief bill that provides a corporate tax cut for domestic manufacturing operations. I am pushing to get small manufacturers (S-corps, LLCs, sole proprietorships, etc.) who do not pay corporate taxes included in the legislation before it is sent to the president.

Manufacturers say the cost of doing business in the U.S. is so high because of healthcare costs, regulatory costs, etc. Is there any relief the U.S. government can give?

As I mentioned earlier, we have been working on an agenda to make U.S. companies more competitive by lowering the cost of doing business here. The main elements of this agenda involve tax relief, healthcare reform, regulatory relief, trade fairness, lifelong learning, spurring innovation, energy self-sufficiency, and ending lawsuit abuse.

The House has already passed several pieces of legislation this year that would dramatically reduce the cost of doing business in the United States. Bills to reform our out-of-control medical liability system and to allow national associations to offer health insurance at group rates to members have already passed the House and are currently awaiting Senate action.

The House also passed an energy bill as well as legislation to extend higher Section 179 expensing limits for capital purchases which have prompted many companies to expand and hire new workers.

What are some of the positive things you can point to regarding manufacturing?

After several years of job losses, the manufacturing sector of our economy is finally starting to create jobs. In addition, the manufacturing employment survey reached a 30-year high in May and remained strong in June, indicating strong future gains in manufacturing jobs.

Also, I believe our nation's political and business leaders are much more concerned about the future of manufacturing in America and are taking steps to strengthen this critical sector of our economy. When I formed the House Manufacturing Caucus a year ago, we had 29 members of Congress. Today, we have more than 80 members. As chairman of the House Committee on Small Business, I have held more than 60 hearings on the importance of manufacturing in America. I believe our nation's leaders now realize how crucial manufacturing is, not only to our economy but to our national security.

If you could only say one thing to manufacturers, what would it be?

I would say, "Don't get too comfortable, and stay on the cutting edge." While the worst times appear to be over and business is picking up for many, you never know when the next hard times are around the corner. I have visited hundreds of manufacturing facilities over the years, and I can tell you the ones that stayed innovative and continued to search for new markets—even during the good times—were the ones who survived the recent recession. Prosperity tends to breed contentment. You need to maintain your edge.

How can we make vocational opportunities in manufacturing seem appealing to America's youth? What do you think needs to be done?

I believe one of the main reasons manufacturing has lost its prominence as a preferred career field dates back several decades when our leaders began promoting college as the only avenue for success in young people's lives. Not every young person was meant to go to college, and they were set up for failure if they did not receive a degree.

I believe we are making progress in changing this attitude. The House Republican Conference adopted "lifelong learning" as a critical component of our competitiveness agenda, and the House recently passed legislation to provide tax credits to companies who send their employees to school to get the latest advanced-technology training. As you know, manufacturing is becoming more and more technical and our companies' abilities to continue to compete will depend on a well-educated workforce.

We must also encourage our high schools to continue to offer vocational opportunities to their students. I was honored recently to give the closing speech at a "manufacturing camp" in Rockford, Ill., which spent a week teaching 8th, 9th and 10th graders the concepts of manufacturing. We need more opportunities like this for our young people.

MARCY KAPTUR
Congresswoman (D-Ohio)

Rep. Marcy Kaptur (D-Ohio) represents the Toledo area's Ninth Congressional District in Northwest Ohio and is serving her 11th term in the U.S. House of Representatives. She is the senior-most Democratic woman in Congress and ranks as the senior Democratic woman on the exclusive House Appropriations Committee. She is one of only 76 women out of 535 members of the 108th Congress.

Kaptur has consistently fought against what she considers unfair trade agreements like NAFTA and GATT and most-favored-nation status for China, stating they are unfair to American families, workers, and businesses.

Representative Kaptur's family operated a small grocery where her mother worked after serving on the original organizing committee of an auto trade union at Champion Spark Plug. She became the first family member to attend college, receiving a scholarship for her undergraduate work. Trained as a city and regional planner, she practiced 15 years in Toledo and throughout the United States before seeking office. Appointed as an urban advisor to the Carter White House, she maneuvered 17 housing and neighborhood revitalization bills through the Congress during those years.

Subsequently, while pursuing a doctorate in urban planning and development finance at the Massachusetts Institute of Technology, her local party recruited her to run for the House seat in 1982. She had been a well-known party activist and volunteer since age 13. Though outspent by 3 to 1 in the first campaign, her deep roots in the blue-collar neighborhoods and rural areas of the district made her race the national upset of 1982.

Kaptur fought vigorously to win a seat on the House Appropriations Committee. Since elected, she has risen in seniority and is now the senior Democratic woman on Appropriations. She has secured subcommittees on Agriculture, the leading industry in her state, and Housing and Urban Development, Environmental Protection, Veterans, NASA, and the National Science Foundation, which allow her to pursue her strong interests in economic growth and new technology, community rebuilding, and veterans. In her legislative career, she has also served on the Budget; Banking, Finance and Urban Affairs; and Veterans Affairs committees.

What are your thoughts on NAFTA? Is it working? Why or why not?

NAFTA aimed at continental "free trade," i.e., tariff elimination, between the United States, Mexico, and Canada.

Yet by the early 1990s, most tariffs already had been reduced between the three nations, with an effective overall tariff rate of about 2%.

Indeed, NAFTA concerned something else. Its unstated aim was to provide a government sanctioned insurance scheme for rising investments by transnational corporations in low-wage nations starting with Mexico, which was close to the U.S. market, and where subsistence labor was plentiful. NAFTA accelerated the shipping out of U.S. jobs.

The populations of Mexico and Canada totaled more than 125 million persons. Mexico's largely poor population equals more than 100 million and its workers are fearful about organizing trade unions to gain living wages. The low-wage pull was irresistible. By the early 1990s, the United States was already falling behind Europe and Asia as its global trade deficit in goods rose with each passing year.

With NAFTA's passage, the export of U.S. jobs to Mexico exploded. Mexico started to import vast quantities of Chinese products that then backdoored their way into the U.S. The United States' job market began to shift millions of jobs to Third World environments as reflected in rising global trade deficits. Outsourcing of production and services, even of American icon products like Amana, Brach's, Hoover, and the PT Cruiser, became commonplace and accelerated.

Outsourcing has received much press lately. What are your opinions on manufacturers outsourcing? Is all outsourcing bad, or is it just inevitable?

Outsourcing is a last resort. It sounds like we are getting extra help, but what it really means is since we no longer have manufactured goods to export, we will now export the last thing we have left—our jobs.

It isn't inevitable. It is surrender. When there are skilled Americans looking for work, our first priority must be to help them find meaningful employment. We know from our experiences with NAFTA that trade agreements are often excuses to find cheap labor, few demands on manufacturing practices, and any other device for improving the "bottomline." What is the bottomline for our national economy?

When we have to spend money on unemployment assistance we will do so because American workers deserve help. Every unemployed person with whom I have spoken would far rather have a job than this assistance. When my constituents learned that even the help lines for programs like food stamps were being sent overseas, they were horrified. That is why the House adopted my amendment to prohibit federal funds from being used to pay the charges for call centers sent overseas. Having federal dollars paying for work sent overseas is just as wrong as having tax breaks that encourage companies to build plants overseas.

What are your thoughts on protecting the Intellectual Property rights of American manufacturers?

Across our country we see the dismantling of jobs and business in this country. If one reads Article 1, Section 8 of the U.S. Constitution, it says, the Congress shall have the power to secure for inventors the exclusive right to their respective writings and discoveries. Throughout the more than 200-year history of our country, that has been done through the U.S. Patent and Trademark Office.

The WTO and other trade regimes seek to strip inventors of the right to protect their inventions and intellectual property. I have spent years fighting for the right to patent and protect intellectual property. It is one of our nation's greatest strengths that must be protected at all costs.

I have introduced a bill that would impose a tariff on products imported to the United States that are derived from varieties of seed for which foreign producers have not paid the patent royalty that is charged to American producers. We should consider the same policy for all imported products that violate our intellectual property laws. We should not allow our patent protections to be circumvented.

There is a lot of discussion about free trade vs. fair trade? How would you respond to manufacturers about the difference and what you believe needs to be done?

Free trade is only "free" when we are trading in fair manner with free people. American companies and American workers continue to suffer from a massive and growing international trade deficit that is destroying jobs and tearing apart communities. The current trade policy is an abject failure. It has long been a priority for me to help fashion a new American fair trade policy that serves the needs and concerns of working people. Last year, I led a congressional-labor delegation to Mexico to examine the impact of NAFTA after 10 years.

We saw once more that the primary effect of NAFTA has been a shift of production and jobs to a low-wage country without real labor unions or meaningful environmental law enforcement. The "maquiladora" factories near the border turn out goods mainly for the U.S. and European consumer markets, but the workers do not earn enough to buy the very products they make.

The net effect is that Americans lose good-paying jobs, American consumers pay the same price for goods (or even higher prices), and Mexican workers still live in poverty. NAFTA and trade agreements such as the Singapore and Chile measures primarily serve the interests of large corporations and international investors.

I will continue to fight against the expansion of NAFTA. I will continue to fight for fair trade agreements that help raise the standard of living for all people.

10

The Mending Has Begun

While the economy is experiencing a steady boost that has been fueled by government spending practices, tax cuts, mortgage refinancing, and other one-time stimuli, this is only the beginning. To sustain growth manufacturers need to become more active and need to spark new and innovative methods to grow their operations here and abroad. Looking forward there appears to be nothing on the horizon that will continue to stimulate the economy more than a successful drive spearheaded by American manufacturers. It is fair to say that while we are beginning to see some progress in manufacturing, we are far from where we need to be to remain a viable player in world markets. Thus, American producers need to become more involved in their future and the destiny of their manufacturing operations.

So what makes a manufacturer optimistic these days? We have heard it a million times before that small and medium-size manufacturers are less than enamored with the current market conditions. Despite economists' assurances about the broad efficiencies and benefits of global commerce, they are not bringing solace to SMMs who have spent a lifetime building and refining their company only to watch its business disappear right before their eyes. And to make matters worse, some manufacturers—who have built their companies from the ground up—insist that today's global commerce is crippling them in ways they have never experienced before. Even though many manufacturers feel battered and bruised, it is important to remember that global commerce does create more opportunities, such as lower prices, a possible increase for exports, and more spending power.

In a highly competitive landscape, heading up a manufacturing company is certainly no easy task. Toss in today's economic environment, global competition, and thinning margins, and manufacturers are working more than just overtime to keep their plants going. Running a manufacturing company has always required a solid commitment, a strong constitution, a remarkable resiliency, and an uncanny desire to succeed at all costs. Today, these qualities are merely a starting point. Today's top executives also need to be so innovative that they not only lead the charge to dominate a market, but they are always striving to stay ahead of the competition here and overseas.

After nearly three years of slumping production numbers, increasing layoffs, and continued outsourcing of jobs, just how long does it take to add any degree of encouragement to the industry? According to the ISM it takes about nine months. Its purchasing-manager index, which is a survey of supply executives, was above 50—the dividing line between industry growth and contraction—for the 13th consecutive month in June. And according to survey chief Norbert Ore, because of this figure, coupled with the fact that 17 industries within manufacturing reported growth for the month of June, the sector clearly has sustainable momentum at this point.

Or does it? While the numbers clearly indicate the industry is gaining momentum, for a great many manufacturers in this country it is going to take much more than a PMI number above 50 for the next several months to convince them that manufacturing isn't still lingering in dire straits.

Most manufacturers are very familiar with the facts and figures surrounding manufacturing in this country. The bigger question lies in whether manufacturers can really understand and interpret the numbers to be in their favor.

Before we can look ahead let's look back at some of the predictions made about the economy. The U.S. treasury secretary predicted U.S. economic growth could reach as high as 4% by mid-2003. He cited support for this prognosis by repeating that the U.S. economy grew an average of 3% in the first nine months of calendar 2002, when in reality the economic growth for 2003 hit 4.2%.[1]

Even The International Monetary Fund (IMF), adjusting one forecast after another downward, set U.S. economic growth in 2002 at 2.2%, down from a previous forecast of 2.3%, and pre-

Figure 10.1 **U.S. ECONOMIC GROWTH**

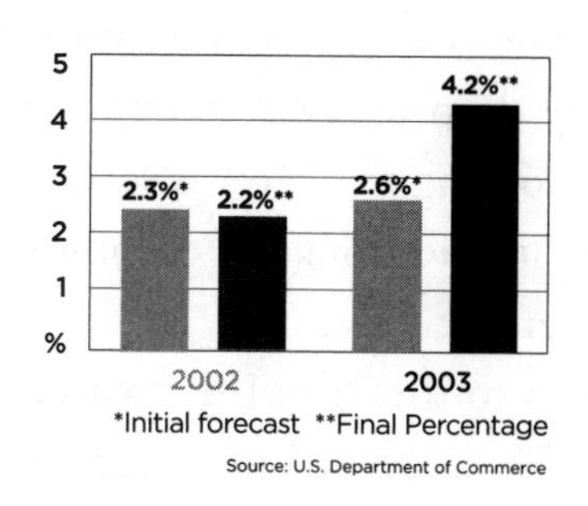

*Initial forecast **Final Percentage

Source: U.S. Department of Commerce

dicted it would likely grow 2.6% in 2003, down from a previous estimate of 3.4%. In actuality, as noted earlier, economic growth for 2003 was 4.2%.[2]

The Chicago Federal Reserve Bank, another economic prognosticator, estimated that U.S. economic growth was likely to accelerate in 2003, reaching close to a 3.5% annual growth rate. The Chicago Fed was only off slightly less than one percentage point. The Fed's 0.5 percentage point interest rate cut on Nov. 6, 2003 to a 41-year low of 1.25% was critical in bolstering demand and allowing U.S. growth to pick up in 2003.[3]

Merrill Lynch predicted the U.S. economy would grow 3.3% in 2003, compared with a projected rate of 2.6% in 2002. The investment banker said the growth rate would stand at 2.5% in the first quarter of 2003, 4% in the second quarter, and 4.5% in the third and fourth quarters. Merrill Lynch was correct on most of the quarters except the third, which saw an incredible 8.2% growth.

The Korea Center for International Finance (KCIF) was among the closest prognosticators predicting the U.S. economy would grow 3-4% in 2003, up from the 2002 estimate of 2.5%. In a global economic report, KCIF predicted the U.S. to show a modest economic recovery in the latter part of 2003 (which it had), boosted by an anticipated rise in labor productivity and facility investment.[4]

The center also said unit labor costs in the U.S. would likely drop because of improving labor productivity, while the housing market was sounding an upbeat note with positive implications for economic growth. All of these have proven accurate.

As to the rest of the world, it is encouraged by a better pace of recovery in Canada, Mexico, and Chile, and especially by the prospect of rapid expansion in Asia where growth in gross domestic product remains strong in China and has picked up in Korea, Hong Kong, Taiwan, and Malaysia.

The center expected China, the world's fastest growing economy, would have economic growth of more than 7% in 2003, bolstered by rising exports and fixed investments. However, China faces continuing problems of a budget deficit and increasing bad loans at banks. However, despite the booming markets overseas, the center has warned that Asian countries might be hit by deflation, which is said to be already happening in Japan, China, and Hong Kong.

Thus, if all of these reports, industry observers, and the Fed are correct, manufacturers should, and will be, the beneficiary of an economic recovery that is long overdue. The real question is whether the United States needs to focus on managing its own behavior and controlling access to its own markets before real sustained growth will occur.

Therefore, the facts: manufacturing makes up one-fifth of the gross domestic product, it is responsible for approximately 15 million jobs in the United States, it supports another eight million jobs in other sectors and it accounts for 62% of all research and development.[5]

We are all equally familiar with the other end of the spectrum, which states manufacturing has lost nearly 3 million jobs since the end of 2000. In addition, even more manufacturers are outsourcing much of what used to be built and assembled right here in the United States.[6]

Understanding these numbers and then evaluating them will help manufacturers to better prepare and plan for the future. Rather than fighting the rapid changes in globalization, American manufacturers need to embrace these markets.

With all the negative news and devastating numbers that are released almost daily, the majority of American manufacturers are not on the brink of throwing up their hands in defeat. On the contrary, if anything, manufacturing has been built on perseverance and prosperity and there are some who perceive this latest setback as only another hurdle in their goal for long-term sustainability.

Small and midsize manufacturers, which have long been the backbone of the domestic manufacturing base, are pros when it comes to adapting to adversity and have taken the current economic crisis in the same fashion by addressing new markets, pursuing new customers, or revamping their entire go-to-market strategy altogether. Yet, even with new strategies, many believe the United States government has abandoned them.

It's obvious to those who really understand manufacturing that not only the administration, but all of our government officials, cannot forget that this country was built and continues to prosper because we have the greatest manufacturing in the world. Without a strong manufacturing sector, the United States could become a Third World country. We need to heed the warning, however, that we cannot become a service-oriented-only country. And unless we are all willing to take corrective measures, America could become a second-class manufacturing power. However, with the passion and commitment of government officials, the administration, and all Americans alike, we can overcome.

As manufacturers entered the second half of 2004 they appeared to be filled with cautious optimism. While many are optimistic about the immediate future, they are pessimistic about the fact that the industrial sector does not have the government encouragement to do more manufacturing in this country. Therefore, manufacturers are seeking a government that works with manufacturing to help it become productive.

Many admit their frustration is aimed at the administration and other elected officials who are not doing more to remedy the problems in manufacturing. Many insist the government needs to lend more of a helping hand to the small and medium-size manufacturers in this country. While the government can do more to help mend manufacturing, Americans need to do even more to rectify the downward spiral.

Unfortunately, waiting for a clear answer to the problem facing American manufacturing may be a long and fruitless endeavor. While many have adapted to suit the changing nature of the situation, the question becomes, what will it take to invoke a permanent turnaround?

It is a world economy today and companies have to decide that if they are going to continue to be manufacturers they had better plan on doing some business outside of the U.S., but if they are a smaller company they need to work on providing a product or service that their customers cannot find anywhere else, and do things that others simply cannot do.

To address this notion, many other manufacturers are rallying together at manufacturing summits all over the country to create a unified voice to let the government know that without a strong manufacturing base, the future of the country is in serious trouble.

Many of the manufacturers who attend these summits are mainly small and midmarket companies, the very businesses that have been affected the most by this industrial downturn.

Manufacturers themselves are also taking center stage by publicly sharing their manufacturing problems. For instance, one Florida manufacturer has begun to share his story of how some manufacturers have to conduct business in an unfair economic climate. This manufacturer was a victim of intellectual property theft. His designs and documents were stolen by a Chinese manufacturer who was counterfeiting the U.S. manufacturer's products overseas.

Another example of a company doing whatever it takes to remain competitive despite intense pressure from foreign competition comes from a major manufacturer of office furniture. The U.S. furniture-making industry remains among the largest segments battling against cheaper products from overseas competition. However, this furniture maker decided to be proactive and began purchasing small amounts of these imports and reselling it. This enabled them to compete in the low-end furniture market allowing their name to remain at the forefront of the market in the U.S. While the company realizes its attempt to thwart foreign imports from entering its market segment is a short-term solution, the company believes this is enabling it to keep its supply chain intact. Another is a metal stamp manufacturer who has voiced concerns about his company's bold step to secure its future by cutting off $1 million worth of business so that it can focus on other markets that could eventually bring double-digit growth.

Another is a maker of robotic arms who has publicly shared eye-opening presentations about the amount of waste that the federal government generates, money, he says, that could be better spent to help the backbone of the U.S. economy: SMMs.

The paper industry, yet another example, is always

changing as manufacturers come up with different ways to package their products, which typically requires changes to their plant-floor operations, which in turn means money spent on capital equipment such as machines and automation equipment.

Manufacturers are taking a bold stance of communicating their concerns and pressures in an attempt to help educate American consumers about the issues that are hindering the strength of the U.S. economy.

As the success stories continue to emerge and the smaller voices continue to band together to form one echoing voice throughout the industry, the small headway made today, say some manufacturers, could lead to large inroads for the future prosperity in American manufacturing.

If anything, manufacturers across the country have not taken the issue of losing jobs to overseas competition lying down. From starting up local manufacturing alliances to attending conferences that address the needs of the market, many manufacturers are ensuring their voice is heard in Washington.

During the past several months various organizations have created several events to bring manufacturers together. The goal is for attendees to network to foster the exchange of ideas in an effort to rebuild American manufacturing, while creating a new circle of life for manufacturers nationwide. The industry needs to discuss what it will take to make manufacturing more viable.

The Manufacturing and Economic Recovery Conference (MERC), hosted by *Start* magazine in early February 2004 and another in September is a prime outlet that brings together manufacturers from across the country to discuss their pains and concerns as one voice about the issues facing

American manufacturing and to map out solutions that can be brought before the powers-that-be in Washington.

While manufacturers are surviving short-term, what most consumers do not realize is that workforce reductions and such will allow companies to survive in the short-term only. This thinking is short-sighted and it doesn't address the long-term problems that will affect future generations.

Others believe it will take more than just one voice to make a change. Although most manufacturers are cautiously optimistic that the United States will take the necessary steps to protect American workers, many believe that elected officials will not react quickly enough in this election year and it will be too late to help.

Many in the industry believe the manufacturing sector, in essence, can look no further than itself in trying to find out what to do in order to better compete and make American manufacturing a vibrant industrial powerhouse once again. Manufacturers need to do more to help themselves prosper.

Educating the public and the politicians on the actual pains that are occurring and how vital manufacturing is to this country needs to start with the men and women who are going through it on a daily basis.

While much of this book has tried to point out the long-term impact on the United States should manufacturing continue to decline, it's vital to understand that manufacturing productivity will not improve unless we as a country change collectively and respond as one voice.

In order to drive the much-needed change that must occur, the United States must take a closer look at the strengths and accomplishments of U.S. manufacturing, the unique challenges confronting manufacturers today, the politicians who are voting on laws that impact manufacturers, and the legislative mandates necessary to address these challenges.

It can't be stated enough: manufacturing faces unprecedented challenges. One could even liken today's manufacturing emergency to be no-less threatening to America's long-term viability of the industrial sector overall. It is unlikely that manufacturing will continue to enjoy higher productivity rates that it once did and it certainly will not experience the same productivity rates as the rest of the economy. The end result is a continuing long-term decline in the share of jobs in manufacturing and it is likely to continue unless drastic measures are implemented.

Figure 10.2 **EXPORTS** 2002

$125 Billion*

$22 Billion

U.S.

China

*goods sent to U.S. only

While we cannot go back and rectify the problems that have occurred, Washington can slow the rate of decline, particularly if Congress focuses on smart policies to help manufacturing become more competitive. This will include saving jobs that otherwise would be lost to foreign competitors as well as reinvesting in technology.

The strength of the U.S. economy continues to be weakened by the increase in the trade deficit. The bill we run up every year by purchasing more imports than selling exports will have to be paid eventually as foreign nations demand payment in real goods and services. The United States' ability to pay its debt is hindered by its inability to sell more goods in America and abroad. There is nothing reciprocal in a trade relationship where American manufacturers could only sell $22 billion in exports in 2002, while Chinese producers sent to the United States $125 billion of imports, this is nearly 5-1 advantage for Beijing.

If the administration does not act, we could find ourselves in a financial spiral we cannot stop. This financial impact could dramatically increase the national debt and the Social Security shortfall at the same time. The longer we hold off in repaying the U.S. debt could mean a larger devaluation of the dollar, which will inevitably lower the purchasing power for the next generation.

Only a strong collective voice, together with allied public interest groups, can bring about necessary changes that draw the attention of the public, administration, and other elected officials. The ability to succeed depends on all of us working together toward a common goal. Manufacturers, both large and small, need to work together to rebuild American manufacturing.

As this book has revealed, there seem to be many reasons, not just one, for the manufacturing crisis.

Despite these problems, one fact that speaks volumes about the manufacturing industry is its ability to persevere. Now is the time to focus our attention on rebuilding and mending American manufacturing. It's not about finding fault. It's time to come together. It's time to hear the rallying cry of the industry.

One way to solve these problems is to create a true manufacturing community. It's up to each of us to work together to rebuild the economy. The first place we need to begin is to look within the manufacturing industry. It's time to develop a sustainable plan.

Bringing brilliant minds together is just the first step. The next step will be to set a strategic approach. This entails preparing for growth now, so when the growth comes the industry is able to take advantage of it. In addition, that means looking at technology solutions that can reduce manual labor as well as focusing on manufacturing practices. It means diversifying the product mix. It also means investing in R&D. It means developing a strategy that can be executed, and providing technology tools that make employees' jobs easier.

In order to help companies evaluate their markets, *Start* magazine, with the help of other individuals and companies who are committed to revitalizing American manufacturing, has created the Making It Together Alliance.

This alliance is designed to help hard-working individuals discover new ways of networking in the hopes of exchanging new and creative ideas. The ultimate goal is to build new friendships that will ultimately build a stronger tomorrow. With your help and support we can make a stronger tomorrow.

Even during the toughest time in manufacturing history, manufacturers have weathered many horrific storms, and in some cases prospered, either through acquisition, changing their product mix, or even entering new markets.

Despite today's economic environment, SMMs are coming together to uncover new ways to get their plants humming again. The industry is really starting to understand that to be heard in Congress, especially in an election year, it needs a larger voice. It needs one voice. The industry needs to work together toward a common goal.

Alongside of the industry seeking change, it must promote the change to consumers who need to understand that if we do not help spur the manufacturing industry it will not only impact our children, but our children's children. As business owners and leaders of your corporations you are in the perfect position to formalize a competitive plan to help manufacturers across the country.

It's time consumers are made aware that the weakening of American manufacturing only weakens the strength of the United States and its economic health. We Employ America is one key initiative that is designed to spark awareness at a national level among manufacturers and consumers alike. The We Employ America campaign is a new initiative that was launched this year that identifies the companies that make, sell, and service at least 65% of their products within the United States. Unlike the "Made in the U.S.A." mark that was popularized many years ago, the We Employ America program requires manufacturers to prove that a majority of their products are produced within the United States and meet other established criteria. Initiatives like these go a long way in raising consumer awareness about the products they are purchasing and how their retail decisions impact the economic well-being of the United States.

If all Americans work together the manufacturing industry will continue to gain momentum to drive growth.

This alliance is designed to help hard-working individuals discover new ways of networking in the hopes of exchanging new and creative ideas. The ultimate goal is to build new friendships that will ultimately build a stronger tomorrow. With your help and support we can make a stronger tomorrow.

In sum, the challenges facing American manufacturers will continue for years to come. As a result, now is the perfect time to join forces to forge even greater pathways to competitiveness. Manufacturers of all sizes and shapes need to reinvent themselves and to take stock of what their companies make today and what they will make tomorrow. As Americans we cannot forget that manufacturing is the heart of our economic strength and our national security. We cannot maintain our leadership position among the family of nations without a strong and viable industrial sector.

There is no better time to raise awareness of how vital manufacturing is to the economic well-being of the United States. In addition, we need to identify ways to be more competitive while eliminating the impediments that are hindering the prosperity of American manufacturing. If and when we all come together we can mend American manufacturing.

References

CHAPTER ONE

1 U.S. Dept. of Commerce, "Manufacturing in America: A Comprehensive Strategy to Address the Challenges to U.S. Manufacturers," http://www.manufacturing.gov/mfg_complete_low_res.pdf, January 2004.

2 United States Business and industry Council (USBIC), http://www.americaneconomicalert.org, Washington, D.C., 2004.

3 Estimated based on data from the U.S. Census Bureau, www.census.gov, Washington, D.C., 2004.

4 U.S. Dept. of Labor, www.dol.gov, Washington, D.C., 2004.

5 U.S. Dept. of Labor, www.dol.gov, Washington, D.C., "U.S. Productivity and Cost Index," 2004.

6 U.S. Dept. of Labor, www.dol.gov, Washington, D.C., 2004.

7 U.S. Dept. of Labor, www.dol.gov, Washington, D.C., 2004.

8 U.S. Census Bureau, "U.S. International Trade in Goods and Services Highlights," Washington, D.C., 2004.

9 International Union, United Automobile, Aerospace and Agricultural Implement Workers of America (UAW), www.uaw.org/cap/02/issue/issue06-2.html, 2002.

10 U.S. Census Bureau, "U.S. International Trade in Goods and Services Highlights," Washington, D.C., 2004.

11 Federal Reserve Bank of San Francisco, The., http://www.frbsf.org/economics/index.html, "Economic Research & Data," San Francisco, Calif., 2004.

12 U.S. Census Bureau, "U.S. International Trade in Goods and Services Highlights," Washington, D.C., 2004.

CHAPTER TWO

1 Institute for Supply Management, "Employment Index," www.napm.org, Tempe, Ariz., 2004.

2 Meckstroth, Daniel. "Economics Update," Manufacturers' Alliance/MAPI, www.mapi.net, Arlington, Va., 2004.

3 Conference Board, The. "U.S. Economic Highlights," www.conference-board.org, New York, N.Y.

4 U.S. Dept. of Labor, "U.S. Import and Export Price Indexes," www.dol.gov, Washington, D.C., 2003.

5 U.S. Dept. of Labor, "U.S. Import and Export Price Indexes," www.dol.gov, Washington, D.C., 2003.

6 Conference Board, The. "Economic Research: The Conference Board," www.conference-board.org, New York, N.Y.

7 JP Morgan Chase & Co. and NTC Research, "Diversified Manufacturing," www.jpmorgan.com, New York, N.Y., and Oxfordshire, England.

8 Institute for Supply Management, "Global Manufacturing Purchasing Manager's Index," www.napm.org, Tempe, Ariz., 2004.

9 Institute for Supply Management, "Global Manufacturing Purchasing Manager's Index," www.napm.org, Tempe, Ariz., 2004.

10 Institute for Supply Management, "Global Manufacturing Employment Index," www.napm.org, Tempe, Ariz., 2004.

11 U.S. Dept. of Commerce, "Manufacturing and Trade Inventories and Sales," www.doc.gov, Washington, D.C., 2002.

CHAPTER THREE

1 "Jobs and Tax Relief Reconciliation Act of 2003," www.whitehouse.gov, Washington, D.C.

CHAPTER FOUR

1 *Start* Magazine, "Coming on Strong: 2004 Start 1,000 Large," www.startmag.com, Carol Stream, Ill., May 2004.

CHAPTER FIVE

1 National Assn. of Manufacturers (NAM), www.nam.org, Washington, D.C.

2 U.S. Dept. of Labor, www.dol.gov, Washington, D.C., "Labor Productivity and Costs."

CHAPTER SIX

1 Harley-Davidson Co., "Company History: From 1903 Until Now," http://www.harley-davidson.com/CO/HIS, Milwaukee, Wis., 2004.

CHAPTER SEVEN

1 van Eeden, Paul, "This Week by Paul van Eeden," www.kitco.com, Champlain, N.Y., 2004.

2 van Eeden, Paul, "This Week by Paul van Eeden," www.kitco.com, Champlain, N.Y., 2004.

3 Dewey Ballantine LLC, "China's Emerging Semiconductor Industry." Semiconductor Industry Assn., http://www.semichips.org/pre_stat.cfm?ID=225, 2003.

4 Semiconductor Industry Assn., "Global Chip Sales Total $12.1 Billion in April 2003," http://www.semichips.org/pre_release.cfm?ID=272, 2003.

CHAPTER EIGHT

CHAPTER NINE

CHAPTER TEN

1 O'Neill, Paul, U.S. Treasury Sec., "Testimony of Treasury Secretary Paul O'Neill before the Senate Finance Committee." U.S. Dept. of Commerce, www.doc.gov, Washington, D.C., http://www.ustreas.gov/press/releases/po981.htm, 2002.

2 U.S. Dept. of Commerce, "Manufacturing in America: A Comprehensive Strategy to Address the Challenges to U.S. Manufacturers," http://www.manufacturing.gov/mfg_complete_low_res.pdf, January 2004.

3 Moskow, Michael, "Economic Research and Data," Federal Reserve Bank of Chicago, www.chicagofed.org, 2002.

4 Korea Center for International Finance, http://www.kcif.or.kr

5 U.S. Dept. of Commerce, "Manufacturing in America: A Comprehensive Strategy to Address the Challenges to U.S. Manufacturers," http://www.manufacturing.gov/mfg_complete_low_res.pdf, January 2004.

6 U.S. Dept. of Labor, www.dol.gov, Washington, D.C., "Contract Labor, Contracting Out," 2004.

Index

Afghanistan 97
Agriculture, U.S. Dept. of 31
Aldonas, Grant 32, 148-161
Alexis PlaySafe 115, 119
Amana 179
American Apparel Manufacturers Assn. 116
American Productivity Design and Equipment 93
Analog Devices 106
Australia 155

Bahrain 155
Boeing Co., The 119
Bolick, Jack 105, 114
Brach's 179
Burlington Industries 106
Bush, George W. 14, 21, 25, 31, 70, 149-151, 158, 160, 172
Business Software Alliance, The 157

Canada 6, 97, 179, 186
Carey, Dwight 93-104
Center for Business and Public Policy 162
Champion Spark Plug 177
Chile 155, 182, 186
China 5, 6, 7, 8, 14, 20, 21, 23, 29, 30, 32, 46, 49, 50, 52, 53, 56, 60, 72, 73, 74, 77, 84-87, 90, 91, 97, 98, 107, 110, 123, 128, 129, 131, 133, 135, 136, 151, 152, 157, 160, 165, 172, 173, 187, 193, 194
Clinton, Bill 24, 25
Coca-Cola 122
Collins, Mike 58, 79, 82
Commerce, U.S. Dept. of 16, 32, 33, 70, 116, 148, 153, 154, 157, 159, 160
Conference Board, The 40, 43
Congress, U.S. 15, 16, 67, 69, 70, 75, 139, 142, 173, 180, 193, 196
Congressional Business Advisory Council 94
Cornell University 128
Costa Rica 155
Customer-Relationship Management 59
Cyprus 42
Czech Republic 42

Dewey Ballantine LLC 52
Duke University 139

Economist, The 164
EMF Corporation 134
English, Phil (R-Pa.) 173
Enterprise Resource Planning 59
Estonia 42
European Union 43, 97
Evans, Donald 17, 70, 148, 153, 157, 159, 172
Federal Reserve, U.S. 28, 96, 187
Ford Motor Co. 152
Foreign and Commercial Services 153
France 151
Frink, Al 160

Gallup Small Business Index 19
GATT 97, 98
General Accounting Office 5
General Motors 97
Germany 107, 151
Gettysburg College 139
Global Manufacturing Employment Index 46
Global Manufacturing Input Price Index 45
Global Manufacturing New Orders Index 45
Global Manufacturing Output Index 45
GOP 16
Greenspan, Alan 16
Gross Domestic Product 39, 40, 86

Harley-Davidson 74
Harvard Business School 119
Honduras 155
Honeywell Process Solutions 105
Honeywell 106
Hong Kong 187
Hoover 179
Hungary 42
Hussein, Saddam 70

Iacocca, Lee 111
IMF 98, 185
India 107, 118, 129, 136

Indiana University 134
Industrial Revolution 29
Institute for Supply Management 37, 39, 41, 44, 151, 184
International Federation of Purchasing Materials Management 44
International Resistive Company 106
International Trade Administration 148
International Trade Commission 74
ISO 59

Japan 45, 46, 97, 187
Johnson Matthey 106
Justice, U.S. Dept. of 156

Kaptur, Marcy (D-Ohio) 177-182
Keenum, Rhonda 160
Kerry, John 14
Kingstone, Brett 122-127
Korea Center for Int'l Finance 186
Korea 107, 133, 187

Labor, U.S. Dept. of 27, 41
Latvia 42
Lautenberg, Frank 162
Lean Manufacturing 81
Lithuania 42

Making It Together Alliance 195
Malaysia 187
Malta 42
Materials Requirements Planning 59
Manufacturing Caucus 88
Manufacturing Execution Systems 59
Manzullo, Donald 88, 171-176
Massachusetts Institute of Technology 178
MEP 67, 152, 154
MERC 191
Merrill Lynch 186
Mexico 6, 49, 50, 98, 136, 151, 179, 181, 186
Microsoft 113
Morocco 7, 155

NAFTA 7, 10, 50, 97, 98
National Assn. of Manufacturers 66
NIST 154
North Carolina A&T State University 106

OPEC 97
Oxfam International 10
Ore, Norbert 185
Oseco 128

Panama 155
Paul, Ron (R-Texas) 88, 139-147
Pharmaceutical Research and Manufacturers Assn. 69
PMI 37, 38, 40, 44, 45, 185
Poland 42
Product-Lifecycle Management 59
Purdue University 134

Race to the Bottom 10
Ragosta, Joseph 128-133
Ramapo College 162
Reagan, Ronald 74, 111
Russia 98
Ryan, Tim (D-Ohio) 88

Sabrin, Murray 162-170
Sanders, Howard 134-137
Semiconductor Industry Assn. 52
Simon, William 139
Singapore 155, 182
Six Sigma 81
Slovakia 42
Snow, John 172
South African Customs Union 155
Specialty Publishing Co. 94
Supplier-Relationship Management 59
Supply-Chain Management 59, 81
Stanford University 122
Start Magazine 2, 58, 59, 76, 94, 191
Super Vision International 122

Taiwan 107, 187
Textile/Clothing Technology Co. 116
Thailand 155
Toro 157
Toyota 97
Total-Quality Management 59
Tri State University 134
United Kingdom 45, 46, 86
United Merchants and Manufacturing 106
U.S. Business and Industry Council 17
Universal Studios 122
Walt Disney 122
Warren Featherbone 115
We Employ America 196
Wells Fargo 19
Western Carolina State University 106
Whalen, Charles E. "Gus" 115-121
WTO 10, 11, 53, 90, 98